T0202892

Communications in Computer and Information Science   1986

Rationale
The CCIS series is devoted to the publication of proceedings of computer science conferences. Its aim is to efficiently disseminate original research results in informatics in printed and electronic form. While the focus is on publication of peer-reviewed full papers presenting mature work, inclusion of reviewed short papers reporting on work in progress is welcome, too. Besides globally relevant meetings with internationally representative program committees guaranteeing a strict peer-reviewing and paper selection process, conferences run by societies or of high regional or national relevance are also considered for publication.

Topics
The topical scope of CCIS spans the entire spectrum of informatics ranging from foundational topics in the theory of computing to information and communications science and technology and a broad variety of interdisciplinary application fields.

Information for Volume Editors and Authors
Publication in CCIS is free of charge. No royalties are paid, however, we offer registered conference participants temporary free access to the online version of the conference proceedings on SpringerLink (http://link.springer.com) by means of an http referrer from the conference website and/or a number of complimentary printed copies, as specified in the official acceptance email of the event.

CCIS proceedings can be published in time for distribution at conferences or as postproceedings, and delivered in the form of printed books and/or electronically as USBs and/or e-content licenses for accessing proceedings at SpringerLink. Furthermore, CCIS proceedings are included in the CCIS electronic book series hosted in the SpringerLink digital library at http://link.springer.com/bookseries/7899. Conferences publishing in CCIS are allowed to use Online Conference Service (OCS) for managing the whole proceedings lifecycle (from submission and reviewing to preparing for publication) free of charge.

Publication process
The language of publication is exclusively English. Authors publishing in CCIS have to sign the Springer CCIS copyright transfer form, however, they are free to use their material published in CCIS for substantially changed, more elaborate subsequent publications elsewhere. For the preparation of the camera-ready papers/files, authors have to strictly adhere to the Springer CCIS Authors' Instructions and are strongly encouraged to use the CCIS LaTeX style files or templates.

Abstracting/Indexing
CCIS is abstracted/indexed in DBLP, Google Scholar, EI-Compendex, Mathematical Reviews, SCImago, Scopus. CCIS volumes are also submitted for the inclusion in ISI Proceedings.

How to start
To start the evaluation of your proposal for inclusion in the CCIS series, please send an e-mail to ccis@springer.com.

Vladimir Jordan · Ilya Tarasov · Ella Shurina ·
Nikolay Filimonov · Vladimir A. Faerman
Editors

High-Performance Computing Systems and Technologies in Scientific Research, Automation of Control and Production

13th International Conference, HPCST 2023
Barnaul, Russia, May 19–20, 2023
Revised Selected Papers

Springer

Editors
Vladimir Jordan (iD)
Altai State University
Barnaul, Russia

Ilya Tarasov (iD)
MIREA - Russian Technological University
Moscow, Russia

Ella Shurina (iD)
Novosibirsk State Technical University
Novosibirsk, Russia

Nikolay Filimonov (iD)
Lomonosov Moscow State University
Moscow, Russia

Vladimir A. Faerman (iD)
Control Systems and Radioelectronics
Tomsk State University
Tomsk, Russia

ISSN 1865-0929   ISSN 1865-0937 (electronic)
Communications in Computer and Information Science
ISBN 978-3-031-51056-4   ISBN 978-3-031-51057-1 (eBook)
https://doi.org/10.1007/978-3-031-51057-1

This Springer imprint is published by the registered company Springer Nature Switzerland AG
The registered company address is: Gewerbestrasse 11, 6330 Cham, Switzerland

Paper in this product is recyclable.

Preface

The 13th International Conference on High-Performance Computing Systems and Technologies in Scientific Research, Automation of Control and Production (HPCST 2023) took place at the Altai State University on May 19–20, 2023. Altai State University (AltSU) is in the center of Barnaul city – the capital of the Altai region in the southwestern part of Siberia.

HPCST is a regular scientific meeting that has been held annually since 2011. It attracts specialists in the various fields of modern computer and information science, as well as their applications in the automation of control and production, in mathematical modeling and in computer simulation of processes and phenomena in natural sciences by means of parallel computing. Since last year, a subsection on information security was also established.

The goal of the conference is to present state-of-art approaches and methods for solving contemporary scientific problems and to exchange the latest research results obtained by scientists from both universities and research institutions. All the reported results are valuable contributions to the field of applied information and computer science.

Sessions of the conference are devoted to the relevant scientific topics:

- architecture and design features of high-performance computing systems;
- digital signal processors (DSP) and their applications;
- IP-cores for field-programmable gate arrays (FPGA);
- technologies for distributed computing using multiprocessors;
- GRID-technologies and cloud computing and services;
- high-performance and multiscale predictive computer simulation;
- computing in information security services;
- control automation and mechatronics.

Since the first time the conference was introduced at the international level back in 2017, more than 160 researchers from Russia, China, Ukraine, Kazakhstan, Kyrgyzstan, Uzbekistan, Tajikistan, Vietnam and Brazil have participated in the conference. The average number of participants in a single year was about 60 for a long time. At the peak in 2021 the event was attended virtually or personally by more than 140 scholars.

In the last two years, due to the complex political situation in Russia, the positive internationalization trend was interrupted. So this year the conference was attended virtually or *in presentia* by 74 scholars with 72 accepted reports. The reports were chosen by the program committee from 81 qualified submissions (6 submissions were declined by the editors during the routine entry check). However, four international papers were presented from Kyrgyzstan, Kazakhstan, and China. There were seven international participants in total this year. The geographical distribution of domestic participants remained the same: the Siberian regions and Moscow predominated among the attendees.

The most significant and highest-quality reported studies were thoroughly reviewed and included in this volume. Therefore, fewer than half of the reported papers were invited to these internationally published post-proceedings. The authors of 34 papers accepted the opportunity to participate in this volume. All papers were revised, extended to full-paper format and submitted to a second round of review. After the review, 21 full papers featuring original studies in the field of computing, mathematical simulation, and control science and information security were accepted for publication. Among them, 13 papers were accepted after minor revision, 8 papers were accepted after major revision. No papers were accepted as is and no guest papers were included in the volume this year. Out of 13 rejected papers, 6 were declined after the revision. The acceptance rate for this volume is 28% (about 90% on the stage prior to the conference, about 50% on the post-conference stage, 62% on the final stage). The vast majority of the conference papers (45 research items) that were not included in this volume were published as regular proceedings.

The accepted papers cover such topics as:

- Hardware for High-Performance Computing and Signal Processing;
- Information Technologies and Computer Simulation of Physical Phenomena;
- Computing Technologies in Data Analysis and Decision-Making;
- Information and Computing Technologies in Automation and Control Science;
- Computing Technologies in Information Security Applications.

To select the best papers among those presented at the conference, the following procedure was applied.

1. Session chairs prepared a shortlist of the most significant original reports which had a clear potential to be extended to full-paper format.
2. The editorial board, comprised of the session chairs and corresponding editor, made a list of 35 items.
3. Authors of the selected manuscripts then were contacted and asked to extend their papers and resubmit them for review.

Every paper, with no exceptions, was reviewed by at least three experts. In addition to a routine plagiarism check with iThenticate.com, we applied another check with Antiplagiat.ru. The goal was to detect and decline papers that were already published in Russian. A single-blind review method was applied. The review criteria are below:

1. technical content;
2. originality;
3. clarity;
4. significance;
5. presentation style;
6. ethics.

The organizing committee would like to express our sincere appreciation for the organizational support to the administration of Altai State University and to the staff of Institute of Digital Technologies, Electronics and Physics of Altai State University. Only the outstanding effort of the technical staff made the conference possible in a time of travel restrictions.

The editors would like to express their deep gratitude to the Springer editorial and production teams for the provided opportunity to publish the best papers as post-proceedings and for their great work on this volume.

October 2023

Vladimir Jordan
Ilya Tarasov
Ella Shurina
Nikolay Filimonov
Vladimir Faerman

Organization

General Chair

Vladimir Jordan   Altai State University, Russia

Program Committee Chairs and Section Chairs

Ella Shurina   Novosibirsk State Technical University, Russia
Ilya Tarasov   Russian Technological University, Russia
Nikolay Filimonov   M.V. Lomonosov Moscow State University, Russia
Vladimir Faerman   Tomsk State University of Control Systems and Radioelectronics, Russia

Organizing Committee

Vasiliy Belozerskih   Altai State University, Russia
Alexander Kalachev   Altai State University, Russia
Vladimir Pashnev   Altai State University, Russia
Viktor Sedalischev   Altai State University, Russia
Yana Sergeeva   Altai State University, Russia
Igor Shmakov   Altai State University, Russia
Petr Ulanov   Altai State University, Russia

Program Committee

Viktor Abanin   Biysk Technological Institute, Russia
Valeriy Avramchuk   Tomsk State University of Control Systems and Radioelectronics, Russia
Sergey Beznosyuk   Altai State University, Russia
Alexander Filimonov   MIREA - Russian Technological University, Russia
Pavel Gulyaev   Yugra State University, Russia
Ishembek Kadyrov   Kyrgyz National Agrarian University, Kyrgyzstan
Alexander Kalachev   Altai State University, Russia

Vladimir Khmelev	Polzunov Altai State Technical University, Russia
Shavkat Fazilov	Research Institute for the Development of Digital Technologies and Artificial Intelligence, Uzbekistan
Vladimir Kosarev	Khristianovich Institute of Theoretical and Applied Mechanics SB RAS, Russia
Nomaz Mirzaev	University of Information Technologies n.a. al-Khorezmi, Uzbekistan
Elena Kruchkova	Polzunov Altai State Technical University, Russia
Aleksey Nikitin	Altai State University, Russia
Viktor Polyakov	Altai State University, Russia
Aleksey Yakunin	Polzunov Altai State Technical University, Russia
Dmitriy Potekhin	MIREA - Russian Technological University, Russia
Sergey Pronin	Polzunov Altai State Technical University, Russia
Anatoliy Gulay	Belarus National Technical University, Minsk
Gambar Guluev	Institute of Control Systems of National Academy of Sciences of Azerbajan, Azerbajan
Viktor Sedalischev	Altai State University, Russia
Vitaliy Titov	Southwest State University, Russia
Pedro Filipe do Prado	Federal University of Espirito Santo, Brazil
Bruno Pissinato	Methodist University of Piracicaba, Brazil

External Reviewers

Alexey Saveliev	Tomsk Polytechnic University, Russia
Alexey Tsavnin	Tomsk Polytechnic University, Russia
Anatoliy Gulay	Belarus National Technical University, Minsk
Andrey Kutyshkin	Yugra State University, Russia
Andrey Malchukov	Tomsk State University of Control Systems and Radioelectronics, Russia
Andrey Russkov	Yandex, Russia
Anton Konev	Tomsk State University of Control Systems and Radioelectronics, Russia
Anton Yants	Perm National Research Polytechnic University, Russia
Bibigul Koshoeva	Razzakov Kyrgyz State Technical University, Kyrgyzstan
Elena Luneva	Tomsk State University of Control Systems and Radioelectronics, Russia
Eugeniy Kostuchenko	Tomsk State University of Control Systems and Radioelectronics, Russia

Contents

Information and Computing Technologies in Automation and Control Science

Computing Technologies in Information Security Applications

Hardware for High-Performance Computing and Signal Processing

Design of a Pipeline Computing Module as Part of a Specialized VLSI

Ilya Tarasov$^{(\boxtimes)}$, Daniil Lyulyava, Nikita Duksin, and Ilona Duksina

MIREA – Russian Technological University, Vernadsky Avenue 78, 119454 Moscow, Russia
tarasov_i@mirea.ru

Abstract. The article discusses the process of designing a computing node based on a synchronous pipelined architecture for operation as part of a specialized VLSI. During the design process, individual pipeline levels generate conflicting requirements for the implementation of individual nodes and the pipeline as a whole. This requires resolution in the form of finding suboptimal pipeline tuning options. The approach discussed in the article involves the use of a high-level synthesizer to distribute calculations between individual stages of the pipeline, which allows to reduce the signal delay at the expense of increased pipeline latency Analyzing the interaction of the pipelined calculator with other components of the computing system makes it possible to include in the model of estimated characteristics the cost of waiting for the result. This method, as the simulation results have shown, significantly corrects the approaches to finding the optimal solution, which also includes the costs of the system resources as a whole. Preliminary estimates have received qualitative confirmation when synthesizing modifications of the test pipeline in CAD FPGA.

Keywords: VLSI · FPGA · Hardware Architecture · Computational Pipeline

1 Introduction

Designing computing systems requires the development of an architectural level and the distribution of tasks between subsystems of various types. A widespread architectural approach is based on the use of general-purpose processor cores [1], including those based on ARM, MIPS, RISC-V architectures [2]. In some cases, the use of microcontroller subsystems makes it possible to solve problems that require high performance if microcontroller cores are used in conjunction with other subsystems [3].

Pipeline computing structures currently act as an effective complement to general-purpose processors when solving a number of computing problems [4]. A number of algorithms can be implemented on the basis of a pipeline with low hardware costs, and the pipelined architecture itself allows flexible adjustment of parameters such as latency, clock frequency and power consumption [5]. However, in the design process, which combines the development of a software model, circuit design at the level of register transfers and the development of a topological representation, conflicting requirements arise for certain nodes. For example, an increase in the frequency entails an increase in the

V. Jordan et al. (Eds.): HPCST 2023, CCIS 1986, pp. 3–15, 2024.
https://doi.org/10.1007/978-3-031-51057-1_1

number of pipeline stages, which negatively affects the delay until the result is obtained [6]. Improving energy efficiency is relevant both for the system as a whole and for individual nodes [7], and technological limitations on heat release density, the so-called "dark silicon effect" [8], play an important role. In such conditions, a natural solution is to search for optimal (or suboptimal) implementation options that take into account the particular quality criteria developed during the design process. The literature notes the importance of problem-oriented approaches, used for the implementation of specialized computing devices, which are nevertheless applicable for fairly mass-produced products [9], not only in the form of processor cores, but also in the form of specialized pipelines [10]. To do this, it is necessary to determine the parameters of the conveyor structure that will be subject to modification, followed by determining the characteristics of frequency, energy consumption and resource utilization.

2   Flow Control in a Pipeline Computing Structure

When placing a pipeline computing structure as part of a VLSI, it is necessary to solve two main problems:

2.1   Development of Components of a Pipeline Computer

In the process of developing components of a pipeline computer, the main difficulty is the optimization of functional units located between the registers separating successive stages of the pipeline. The same transformation can be implemented in many ways, implying the distribution of operations across pipeline stages. Increasing the number of pipeline stages reduces the latency achievable for an individual stage, but this latency should be reduced evenly, avoiding excessive focus on non-critical pipeline stages, which will keep the highest latency constant, which will determine the upper limit of the achievable clock frequency.

2.2   Integration of a Pipeline Computer into a Computer System (VLSI)

Analysis of the interaction of a conveyor computing structure with other components of a computing system allows us to formulate and clarify optimality criteria, as well as threshold values of certain characteristics. In particular, improving computational performance is not always a goal in itself, since other system components may limit the input (and output) data flow to a certain value.

Figure 1 shows the architecture of a pipelined data processing structure that uses the output queue ready signal to control the progress of data along the pipeline.

The considered scheme does not take into account operating scenarios in which the output queue is full, but some stages do not contain current data. Having a pipeline-wide data forwarding signal ("write enable") does not allow data to be advanced at individual stages, which potentially creates the problem of additional latency when freeing up space in the output queue.

However, consideration of typical scenarios for the interaction of a pipeline with a computing system shows that the situation when the output queue is full can be easily

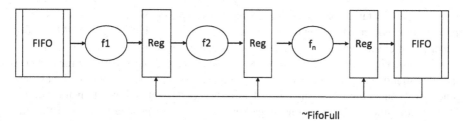

~FifoFull

Fig. 1. Architecture of a pipelined data processing structure using the output queue ready signal.

countered by a corresponding increase in its size in combination with the correct organization of data exchange with the central processor, which can timely read the results of calculations as the queue fills.

3   Selecting the Pipeline Depth Based on Topological Basis

Functional units placed between individual pipeline registers cause a delay in signal propagation between these registers. In this case, the clock frequency of the conveyor will be determined by the maximum value of the signal delay. This means that in order to increase the clock frequency, it is necessary to equalize the delays, redistributing, if necessary, the operations performed at each of the stages of the pipeline.

Increasing the number of pipeline stages has a dual effect. On the one hand, this leads to a decrease in latency at individual stages and a corresponding increase in clock frequency. On the other hand, negative effects include:

1. An increase in the node area by adding additional registers.
2. An increase in power consumption, with an increase in the dynamic component of power consumption occurring both due to increase in clock frequency and due to the addition of registers with high switching activity.
3. An increase in latency when processing a data stream.

It can also be noted that reducing the signal propagation delay in the limit is limited by the uncertainty of delays in the clock network, and when implementing the circuit in an FPGA, also by the maximum size of the combinational circuit of one LUT.

The main relationships in the pipeline circuit are as follows.

The clock period for a pipeline circuit with N stages is given by:

$$t = t_0 + k_n \frac{T}{N}.$$

where k_n – coefficient taking into account the unequal division of the pipeline at the stage ($k_n \geq 1$); t_0 – minimum required time interval, in particular due to the unevenness of the clock signal trace.

The power consumption is determined by static and dynamic components:

$$P = P_{static} + k_{logic}f + k_{reg}Nf,$$

where P_{static} – static component of power consumption, W; k_{logic} – coefficient of the dynamic power component due to switching combinational logic, W/MHz; k_{reg} – coefficient of the dynamic power component due to register switching, W/MHz.

The dynamic component in the above formula is calculated separately for registers and combinational logic. This is due to the fact that the complexity of combinational logic generally remains the same, since it is determined by predefined data transformations. At the same time, adding pipeline stages will sooner or later lead to the fact that the distribution of calculations between them will become unequal, which will lead to a saturation effect for the clock frequency. The number of FPGA logic cells or VLSI gates at which such an effect is observed depends on the selected topological basis and requires separate analysis when moving to a specific hardware platform.

The energy efficiency of a pipeline circuit can be represented as the ratio of clock frequency and power consumption

$$E = f / P.$$

The variety of hardware platforms (both FPGAs and technology library options for VLSI production) does not allow us to clearly indicate the ranges of the given parameters. However, it is possible to consider the impact of the constant latency component on the performance of the pipelined circuit. The importance of this parameter is due to the fact that depending on the platform (FPGA or VLSI) and the category of technological processes (~90 nm or ~28 nm or less), the share of the minimum delay interval increases. This is because the minimum delay value can be 0.2–0.5 ns for clock networks of 28 nm process technology chips (both FPGAs and VLSI), but if the characteristic clock period of the PLL can be in the range of 3–5 ns (i.e., the clock frequency of the project is in the range 200–350 MHz), then for 28 nm VLSI we can count on clock frequencies of 800–1200 MHz, which gives a period of 0.8–1.2 ns. A qualitative view of the dependences of the conveyor clock frequency and power consumption on the number of stages of the conveyor is shown in Fig. 2 and 3 respectively.

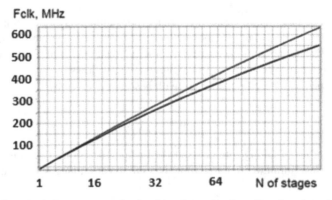

Fig. 2. Pipeline clock frequency vs the number of stages when changing the minimum delay value.

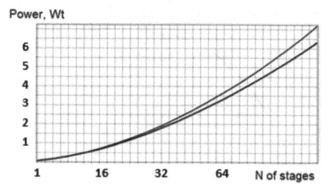

Fig. 3. Power consumption vs the number of pipeline stages when changing the minimum delay value.

When analyzing Fig. 2, 3 it may seem that adding pipeline stages clearly worsens the energy efficiency of the pipeline, since the increase in clock frequency is limited by the minimum signal propagation delay, and adding registers nonlinearly increases power consumption. However, from the point of view of the computing system, reducing the clock frequency means increasing the downtime of components interacting with the computer. Therefore, static power consumption should also include the power consumed by the system while waiting for the result of calculations.

Adding static power leads to the result qualitatively shown in Fig. 4, which shows a graph of the dependence of the energy efficiency of a pipelined computer on the number of stages of the pipeline in the presence of a static component of power consumption. It can be seen that the graph has a pronounced extremum, which allows us to pose the problem of circuit optimization of pipeline computing structures operating as part of specialized VLSI. Static power consumption in this case refers not only to the component consumed directly by the pipeline, but also to the power expended by system components that are forced to stand idle during the operation. This value can be significantly greater than the power of the pipeline, so the optimization problem is determined by the complexity of the interaction of the components of the computing system, of which the designed pipeline is an integral part.

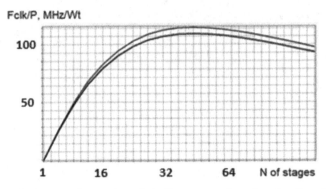

Fig. 4. Energy efficiency vs the number of pipeline stages when changing the minimum delay value.

4   Characteristics of a Pipeline Accelerator in FPGA CAD

To test the selected model, a parameterized pipeline project was implemented, the characteristics of which were assessed in the Xilinx Vivado FPGA CAD system. Since the number of stages of the pipeline was specified by the RTL representation parameter, and the characteristics of the project were determined automatically using a script in the Tcl language built into the CAD system, it turned out to be possible to quickly obtain many pipeline options synthesized using the same CAD settings. An assessment of energy consumption by the Power Estimator utility using a synthesized representation depending on the number of pipeline stages is shown in Fig. 5.

Qualitative view of the graph in Fig. 5 generally confirms the hypothesis about the possibility of optimizing a pipeline computing structure in terms of energy efficiency.

Most accelerators, which are specialized circuits, have a pipeline structure. This is explained by the possibilities for manipulating the main characteristics of the computer to meet the basic technical requirements, which may include the frequency of the circuit, power consumption parameters and other characteristics.

Being an embedded system operating at its own frequency, a specialized computer must have a structure that allows minimal changes to the overall VLSI circuit6 for which it is a submodule. The generalized structure of the computer is shown in Fig. 6.

The presence of input and output buffers is determined by the difference in frequencies between the main part of the VLSI and the specialized computer. The entire pipelined chain is accompanied by a corresponding valid signal chain responsible for forwarding the data. In the simplest case, the chain can provide for stopping the computer when the output buffer is busy (Fig. 7).

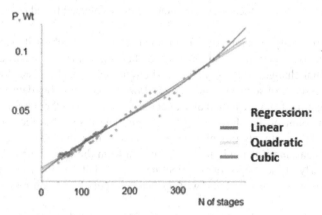

Fig. 5. Dependence of energy consumption on the number of pipeline stages, obtained as a result of a preliminary evaluation, using the Vivado CAD.

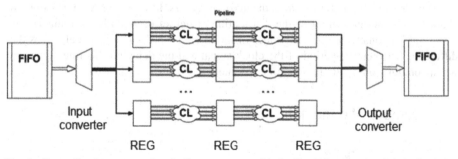

Fig. 6. Generalized structure of a pipeline computer with the interfaces, used to integration into a computing system.

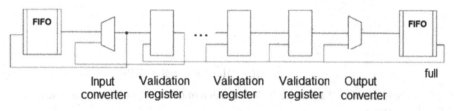

Fig. 7. Flow control of a pipelined computer based on a pipelined data advance signal.

A control subsystem based on a pipelined readiness signal can be accompanied by signals generated according to the principle from the last stage to the first, which subsequently allows you to control possible options for data advancement. At the same time, this approach creates a chain of combinational logic running through the entire pipeline, which has a high chance of becoming a critical path in the pipeline, limiting the clock frequency.

FIFO buffer modules, in turn, must provide the necessary amount of memory for storing information in order to minimize the need to delay data received from the common system bus. The necessary parameters can be selected based on experimental considerations.

Since a pipeline-type computer often uses more data at each of its stages than is implied by the VLSI data bus, and as stated earlier, the computer must be structured to allow minimal changes to the overall VLSI circuit, additional register logic may be required. The essence of such a module comes down to accumulating data for the initial stage of the pipeline from the data that sequentially arrives from the input buffer. When the required amount of data has been accumulated and a signal is received that it is possible to load this data into the initial stage of the pipeline, the module sends the data and sends a signal that it is ready to read new data from the input buffer. The module operates similarly at the output of the pipeline, except that reading from the pipeline is performed in one clock cycle, and writing to the output buffer requires several clock cycles.

As an example, a pipeline based on the CORDIC algorithm is implemented. The project was implemented on the basis of FPGA xc7vx1140tflg1930-1. The results are shown in Table 1.

The slack histogram for the implemented conveyor is shown in Fig. 8. The histogram characterizes the uniformity of the distribution of delays in individual circuits of the circuit and is constructed for the values of the margin time, which is calculated as the difference between the period of the clock signal and the delay on this line under the worst combination of circuit operating parameters.

Fig. 8. Slack histogram for a pipeline with calculation of transcendental functions based on the CORDIC algorithm.

Analysis of the slack histogram shows that most circuits have a uniformly small margin, i.e. the individual stages of the conveyor are implemented quite uniformly. If

Table 1. Results of pipeline implementation for the CORDIC algorithm.

FPGA	Virtex-7 (xc7vx1140tflg1930-1)
Clock frequency	166 MHz
Static power	0.590 W
Synamic power	0.085 W
Worst Negative Slack (WNS)	0.082 ns
Worst Hold Slack (WHS)	0.143 ns
Worst Pulse Width Slack (WPWS)	2.600 ns
Slice LUTs	3106
Slice Registers	1598
F7 Muxes	0
F8 Muxes	0
Slice	830
Block RAM Tile	1
LUT2	2054
FDRE	1530
LUT3	983
CARRY4	743
LUT1	167
FDCE	66
LUT4	41
LUT6	28
LUT5	12
RAM18E1	2
FDPE	2

there are a small group of circuits with a margin of approximately 0, they are the limiting factor, since reducing the clock period will cause those circuits to no longer meet the design timing constraints.

The obtained result showed that the pipeline clock speed for the CORDIC implementation can be increased, provided that the calculation is further distributed across the stages of the pipeline. For FPGAs, the limiting case of pipelining is to use at most one LUT between any two synchronous elements. This approach allows you to get closer to the frequency declared by the manufacturer as the "system clock frequency". For AMD/Xilinx FPGA 7 series, UltraScale, UltraScale+, these values are about 700–750 MHz. A fragment of a pipeline in Elaborated Design mode is shown in Fig. 9.

Since the applied approach to describing the circuit allows us to achieve a high clock frequency, the Xilinx Kria platform, based on an FPGA with 16 nm FinFET technology,

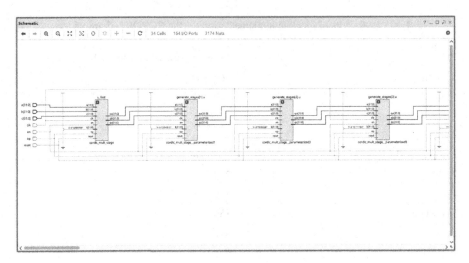

Fig. 9. A part of a CORDIC module in Elaborated Design view.

was chosen as an example. Compared to previous generations, this series has additional latency balancing capabilities by controlling the clock network parameters. This reduces the impact of latency variations caused by not completely identical patterns of the relative arrangement of pipeline stages on the FPGA chip (Fig. 10).

The example above shows that adding topological design constraints makes it easier for the CAD system to achieve high clock speeds. The critical circuits shown in the figure are localized, which demonstrates the dense arrangement of the individual stages of the pipeline. For critical circuits, the share of delay introduced by trace lines is no more than 30% of the total period. Such low performance is generally not typical for modern FPGAs, since the general trend is to increase the proportion of delay introduced by trace lines. If for families of FPGA generations 90–45 nm the ratio of delay on logical resources and trace lines was typically 60/40, then as technological standards decreased and the FPGA area increased, the influence of trace lines increased and currently the recommended ratio is 40/60, whereas in the presented pipeline for worst-case conditions it is 70/30. The slack histogram is shown in Fig. 11.

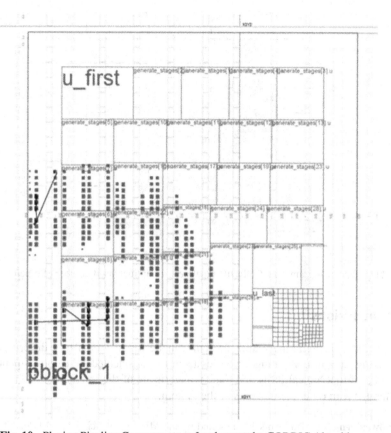

Fig. 10. Placing Pipeline Components to Implement the CORDIC Algorithm.

Given that the period is set to 1.5 ns, the histogram demonstrates the ability to approach the limits of FPGAs when using pipelined computing architectures with balancing of calculations across pipeline stages. It is possible to preserve the uniformly small delays obtained at the synthesis stage by adding topological design constraints and placing logical pipeline cells within the boundaries of one FPGA clock region.

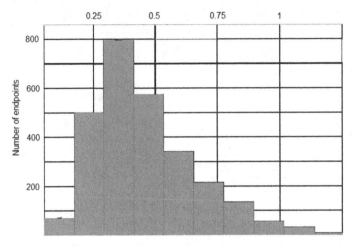

Fig. 11. Slack histogram for a CORDIC pipeline with an increased number of pipeline stages.

5   Conclusions

The results of the preliminary study presented in the article demonstrate the prospects for optimizing pipeline computing structures according to selected indicators, depending on the selected topological basis and the characteristics of the interaction of the pipeline accelerator with other components of the system.

Acknowledgements. This work is supported by the Ministry of Science and Education of RF (Project No. FSFZ-2022-0004).

References

1. Hennessy, J.L., Patterson, D.A.: 2017 Computer Architecture. 6th edn. A Quantitative Approach (The Morgan Kaufmann Series in Computer Architecture and Design) (2017)
2. Harris, S., Harris, D.: Digital Design and Computer Architecture: ARM Edition, 1st edn. Morgan Kaufmann Publishers Inc., San Francisco (2015)
3. Currie, E.H.: Microcontroller subsystems. In: Currie, E.H. (ed.) Mixed-Signal Embedded Systems Design, pp. 35–97. Springer, Cham (2021). https://doi.org/10.1007/978-3-030-703 12-7_2
4. Tarasov, I.E.: Architecture of a specialized VLSI for serial digital signal processing. In: Proceedings of the 3rd International Conference on Control Systems, Mathematical Modeling, Automation and Energy Efficiency, SUMMA 2021, vol. 3, pp. 457–460 (2021)
5. Tarasov, I.E., Potekhin, D.S.: VLSI architecture with a configurable data processing path based on serial distributed arithmetic. In: Journal of Physics: Conference Series, vol. 1615, p. 012001 (2020)
6. Balfour, J., Dally, W.J., Black-Schaffer, D., Parikh, V., Park, J.S.: An energy-efficient processor architecture for embedded systems. IEEE Comput. Archit. Lett. **7**(1), 29–32 (2008)

7. Jhamb, M., Lohani, H.: Design, implementation and performance comparison of multiplier topologies in power-delay space. Eng. Sci. Technol. Int. J. **19**(1), 353–363 (2015). https://doi.org/10.1016/j.jestch.2015.08.006

8. Kanduri, A., Rahmani, A.M., Liljeberg, P., Hemani, A., Jantsch, A., Tenhunen, H.: A perspective on dark silicon. In: Rahmani, A.M., Liljeberg, P., Hemani, A., Jantsch, A., Tenhunen, H. (eds.) The Dark Side of Silicon, pp. 3–20. Springer, Cham (2017). https://doi.org/10.1007/978-3-319-31596-6_1

9. Hennessy, J.L., Patterson, D.A.: A new golden age for computer architecture: domain-specific hardware/software co-design, enhanced security, open instruction sets, and agile chip development. In: Proceedings of 2018 ACM/IEEE 45th Annual International Symposium on Computer Architecture (ISCA), Los Angeles, CA, USA, pp. 27–29 (2018). https://doi.org/10.1109/ISCA.2018.00011

10. Varma, R.A.C., Subbarao, M.V., Varma, D.R., Raju, G.R.L.V.N.S.: High-throughput VLSI architectures for VLSI signal processing. In: Chowdary, P.S.R., Chakravarthy, V.V.S.S.S., Anguera, J., Satapathy, S.C., Bhateja, V. (eds.) Microelectronics, Electromagnetics and Telecommunications. LNEE, vol. 655, pp. 349–358. Springer, Singapore (2021). https://doi.org/10.1007/978-981-15-3828-5_37

Speech Enhancement Based on Two-Stage Neural Network with Structured State Space for Sequence Transformation

Andrey Lependin$^{(\boxtimes)}$ ⓘ, Valentin Karev ⓘ, Rauf Nasretdinov ⓘ, and Ilya Ilyashenko ⓘ

AltSU – Altai State University, Lenin Avenue 61, 656049 Barnaul, Russia
andrey.lependin@gmail.com

Abstract. In this paper, a new method for improving speech quality using the Structured State Space for Sequence (S4) transformation was proposed. This method inherits existing two-stage denoising methods using recurrent neural networks. However, the use of S4 layers instead of long-term short-term memory brought improvements in two ways. Firstly, it was possible to achieve a reduction in the number of trained parameters of the neural network, while maintaining the quality of speech enhancement. Secondly, due to the use of the convolutional representation of S4 transformations, the network training time per one epoch has decreased. The proposed two-stage neural network model for denoising was implemented using the PyTorch library. For training and testing, a standard DNS Challenge 2020 dataset was used. The optimal type of the loss function for training, and the best number of S4 layers was selected. Comparison with existing real-time speech enhancement methods showed that the developed model was one of the best performers for all quality metrics.

Keywords: Speech Technologies · Speech Enhancement · Noise Suppression · Noise Masking · Signal Processing · Deep Learning · Structured State Space for Sequence Model

1 Introduction

One of the most common tasks in modern information systems is the automation of speech processing. Voice interfaces are convenient and widely used when working with conventional personal computers, mobile devices, and smart devices. The quality of the experience of speech technology users is largely determined by the stability of applications and distortions introduced into the recorded speech signal. The development of new effective algorithms for suppressing external noise in an audio signal is critically important.

In many, if not most, practically important tasks of speech enhancement, important requirements for a practically useful algorithm are the high execution speed and the relative simplicity. These requirements are fully satisfied by classical methods for improving the quality of speech (subspace selection, Wiener filtering, and others) [1]. However, their application in modern usage scenarios is very difficult since voice recording is often carried out outdoors or indoors with a lot of non-stationary, high-volume background noise. Under such conditions, modern methods of speech quality improvement based on deep learning technologies work much better [2–7].

The requirement for high execution speed leads to a limitation on the size of the neural net model and the requirement to minimize the delay when processing a noisy signal. Therefore, in most practical noise suppression algorithms based on deep neural networks, the recurrent layers of the LSTM or other types are used [2–5]. In some works [7, 8], attempts are made to use the architecture of transformers applying signal masking.

In the last few years, an effective alternative to recurrent neural network layers has emerged. It is the so-called Structured State Space for Sequence Modeling (S4) transformation [9]. They are successfully applied to the modeling of very long time series when processing text, video, and audio data [9, 10]. They are able to work both in the convolution mode and in the recurrent mode for sequential processing of input data. The convolutional representation of the S4 transformation allows to significantly speed up training a neural network, since an input signal of arbitrary length can be processed as a whole in one step. The recurrent representation allows using a trained "convolutional" network with S4 layers for sequential processing of input data at the inference stage. An extremely important advantage is also the relative simplicity of the structure of the S4 transform itself. The number of trainable parameters in it, as a rule, is less than the recurrent layer similar in terms of capabilities.

In this paper, we propose a neural network model for speech enhancement based on the S4 transformation layers. This model is a modification of the previously proposed model of two-stage speech processing network with recurrent LSTM layers [3]. This modified model has fewer trainable parameters, learns faster and is just effective when used in real time mode.

2 Neural Network Model Used for Speech Enhancement

2.1 Speech Enhancement Applying Complex Ideal Ratio Masks

First, let us describe the formulation of the speech enhancement problem used in this paper. The input signal for the network was represented by the short-time Fourier transform (STFT) coefficients. The input complex spectrogram $\widehat{X}(t, f)$ was considered as the sum of the STFT representations of clean speech $\widehat{S}(t, f)$ and the additive noise $\widehat{N}(t, f)$ [1, 3]:

$$\widehat{X}(t, f) = \widehat{S}(t, f) + \widehat{N}(t, f), \tag{1}$$

where t and f are the discrete time frame number and the frequency component number, respectively. In this work, the influence of the multiplicative distortion was not considered. The STFT spectrogram $\widehat{X}(t, f)$ of the processed signal was fed to the network input. The network predicted a complex ideal ratio mask $\widehat{M}(f, t)$ [11] for the noisy signal in a compressed form (for details see Sect. 3 below). With its help, the complex time-frequency representation of the enhanced signal $\widehat{Y}(t, f)$ was calculated:

$$\widehat{Y} = \left(\widehat{X}_r \odot \widehat{M}_r - \widehat{X}_i \odot \widehat{M}_i\right) + j\left(\widehat{X}_r \odot \widehat{M}_i + \widehat{X}_i \odot \widehat{M}_r\right), \tag{2}$$

where $\widehat{X}_r$ and $\widehat{M}_r$ are the real parts, $\widehat{X}_i$ and $\widehat{M}_i$ are the imaginary parts for the input signal and the mask, j is the imaginary unit ($j^2 = -1$), and $\odot$ is the elementwise multiplication operation.

2.2 Proposed Model

The proposed neural network model (Fig. 1) calculated the output complex mask in two stages, which is conceptually similar to the architecture of the recurrent neural network from [3]. For each discrete time frame t and all frequency components f the vectors of the amplitudes of STFT spectrogram $\widehat{X}(t, f)$ were calculated. They are labeled in Fig. 1 as $\widehat{X}_t$. These vectors were sequentially processed by the full-band block G_{full}. This block was designed to detect patterns in various frequency components of a noisy signal over the entire frequency range. The output of the full-band transformation was the same size as that of the input vector $\widehat{X}_t$:

$$G_{full}\left(\widehat{X}_t\right) = (g(0), g(1), \ldots, g(F - 1)), \tag{3}$$

where F was the number of the processed frequency components.

The S_G transformation divided the output of the G_{full} transformation into overlapping sub-bands for each central frequency f:

$$S_G\left(\widehat{X}_t, f\right) = (g(f - n_G), \ldots, g(f), \ldots, g(f + n_G)). \tag{4}$$

Each sub-band vector $S_G\left(\widehat{X}_t, f\right)$ had a width parameter n_G. For central frequencies f near the boundaries of the frequency range (f < n_G or f > F + n_G − 1) continued cyclically indices were used instead.

The same transformation having the width parameter n_X was applied to the amplitudes $\widehat{X}_t$:

$$S_X\left(\widehat{X}_t, f\right) = \left(\widehat{X}_t(f - n_X), \ldots, \widehat{X}_t(f), \ldots, \widehat{X}_t(f + n_X)\right). \tag{5}$$

The input of the second sub-band processing block (G_{sub} in Fig. 1) was the matrix of the size $(2n_G + 2n_X + 2) \times F$ formed by the concatenated vectors $S_G\left(\widehat{X}_t, f\right)$ and $S_X\left(\widehat{X}_t, f\right)$ for each frequency component f. This matrix represented wide frequency bands of the processed speech signal. Due to this transformation the sub-band processing block could consider connections between multiple formants of the speech. It ensured effective identification of additive sub-band noise components in the audio signal. The result of the sub-band block was the calculated complex mask $\widehat{M}_{t-\tau}$ for the time frame $t - \tau$.

2.3 Structured State Space for a Sequence Layer

The Structured State Space for a Sequence (S4) transformation [9] in the simplest form is defined by two equations:

$$s_k = As_{k-1} + Bu_k, \tag{6}$$

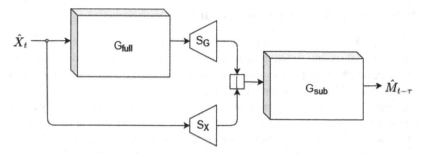

Fig. 1. General scheme of the proposed S4-based speech enhancement model.

$$v_k = Cs_k, \tag{7}$$

where $u_k \in \mathbb{R}^D$ is the k-th vector of the input sequence, $s_{k-1}, s_k \in \mathbb{R}^N$ are the previous and current hidden state vectors of the S4 transformation, $v_k \in \mathbb{R}^E$ is the vector of the output sequence, $A \in \mathbb{R}^{N \times N}$ is the transition matrix for the hidden state and $B \in \mathbb{R}^{N \times D}, C \in \mathbb{R}^{E \times N}$ are two mapping matrices for input-to-hidden and hidden-to-output projections, respectively. This transformation is a discrete analog of a simple continuous dynamical system involving an external control action $u(t)$, an internal state $s(t)$, and an output $v(t)$. Equations (6) and (7) give us recurrent representation of the S4 transformation and could be used for the sequential processing of data.

When training a neural network with S4 layers, it is more practical to use the second, convolutional form of the transformation. It does not depend on the length of the input sequence u_k and is written in the form:

$$V = K * U, \tag{8}$$

where $U = (u_0, u_1, \ldots, u_{L-1})$ and $V = (v_0, v_1, \ldots, v_{L-1})$ are the input and output sequences of the length L, respectively, $*$ is the convolution operation, K is the so-called structured space model (SSM) kernel. The latter is calculated as a result of expanding the chain of the form (for simplicity let the initial state be $s_{-1} = 0$):

$$s_0 = Bu_0, \quad s_1 = ABu_0 + Bu_1, \quad s_2 = A^2Bu_0 + ABu_1 + Bu_2, \ldots$$
$$v_0 = ABu_0, v_1 = CABu_0 + CBu_1, v_2 = CA^2Bu_0 + CABu_1 + CBu_2, \ldots \tag{9}$$

This sequence can be written in a vectorized form:

$$v_k = CA^kBu_0 + CA^{k-1}Bu_1 + \ldots + CABu_{k-1} + Bu_k. \tag{10}$$

For a sequence of the length L, the convolution kernel can then be computed as

$$K = \left(CB, CAB, \ldots, CA^{L-1}B\right). \tag{11}$$

In practice, the direct method for calculating the convolution by Eq. (11) is computationally unstable and time-inefficient. To increase the stability of the learning process, various types of initialization methods are used. The simplest of them is the so-called HiPPO matrix [10], used in this work. To speed up the calculation, the so-called Diagonal Plus Low-Rank SSM is used, which makes it possible to reduce the complexity of calculating the kernel K (for more details, see [12]).

2.4 Full-Band and Sub-band Transform Blocks

When building blocks of full-band and sub-band transformations, recommendations from [13] were taken into account, according to which the most effective scenario for using S4 layers is their alternation with linear layers. In each of the blocks, n repetitions of the linear and S4 transformations were made.

The amplitude vector $\widehat{X}_t$ was normalized in the full-band block (Fig. 2) by dividing the coefficients of this vector by an estimate of the mean value over all previous input vectors at time $0, \ldots, t-1$. The output values of the n structured state space transformation layers were fed to an additional linear transformation followed by the non-linear rectified linear unit (ReLU) function. The arrow inside the S4 layer in Fig. 2 represents the skip-connection in the S4 layer, which is used to improve convergence during training.

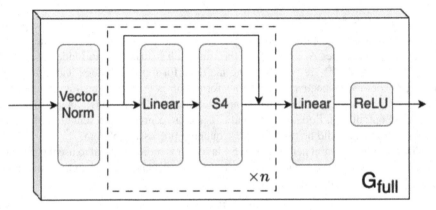

Fig. 2. G-block used for full-band processing.

The sub-band block architecture (Fig. 3) consisted of the normalization of the input matrix of sub-bands followed by n S4 transformation layers. The coefficients of each frequency sub-band (rows of the input matrix) were divided by the estimated mean values of the corresponding sub-band. Unlike the full-band transform, there was no need to use a nonlinear function to limit the linear transformation results for the output mask estimation.

2.5 Loss Functions

To improve the model convergence, the neural network model calculated directly the compressed cIRM mask [11]. Its real and imaginary parts were calculated as follows:

$$\left(\text{cIRM}(\widehat{M})\right)_r = K\left(1 - \exp\left(-C\widehat{M}_r\right)\right)/\left(1 + \exp\left(-C\widehat{M}_r\right)\right), \qquad (12)$$

$$\left(\text{cIRM}(\widehat{M})\right)_i = K\left(1 - \exp\left(-C\widehat{M}_i\right)\right)/\left(1 + \exp\left(-C\widehat{M}_i\right)\right). \qquad (13)$$

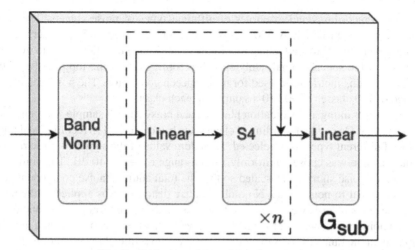

Fig. 3. G-block used for sub-band processing.

The parameters for the compressed mask calculation according to [11] took the values of K = 10 and C = 0.1. For each pair of STFT spectrograms $\widehat{S}(t, f)$ and $\widehat{X}(t, f)$ obtained from the training set, the target complex ratio mask $\widehat{M}^{clean}$ was calculated as

$$\widehat{M}^{clean} = \left(\widehat{X}_r\widehat{S}_r + \widehat{X}_i\widehat{S}_i\right)/\left(\widehat{X}_r^2 + \widehat{X}_i^2\right) + j\left(\widehat{X}_r\widehat{S}_i - \widehat{X}_i\widehat{S}_r\right)/\left(\widehat{X}_r^2 + \widehat{X}_i^2\right). \quad (14)$$

In this paper, three different loss functions [13] were used in the experiments:

1. Mean-square-error (MSE) loss:

$$\mathcal{L}\left(\widehat{M}^{clean}, \widehat{M}\right) = \text{MSE}\left(\text{cIRM}\left(\widehat{M}^{clean}\right), \text{cIRM}\left(\widehat{M}\right)\right). \quad (15)$$

2. Mean-absolute-error (MAE) or L1-loss:

$$\mathcal{L}\left(\widehat{M}^{clean}, \widehat{M}\right) = \text{MAE}\left(\text{cIRM}\left(\widehat{M}^{clean}\right), \text{cIRM}\left(\widehat{M}\right)\right). \quad (16)$$

3. 3. Negation of the signal-invariant signal-distortion ratio:

$$\mathcal{L}(S, Y) = -\text{SI-SDR}\left(\widehat{X}, X\right) = -10\log_{10}\left(\|\alpha S\|^2/\|Y - \alpha S\|^2\right), \quad (17)$$

where $\alpha = Y^T S/\|S\|^2$ is the scale coefficient.

3 Experiments and Results

3.1 Datasets

In this paper, the DNS Challenge 2020 dataset [14] was used to test the effectiveness of the proposed speech enhancement method. This dataset, which has become one of the standard ones in the research area under consideration, consisted of examples of clean speech recordings and examples of different types of noise signals. The clean signals included approximately 500 h of speech recordings from 2150 speakers. These recordings were tested using the mean opinion score (MOS) technique [15] to evaluate the quality of the speech signal. Only speech recordings from the top quartile of the distribution of this metric were used for noisy speech generation. The noise sample set consisted of 150 classes of 500 10-s samples in each class.

In both the training and validation phases, each noisy speech sample was generated in the same way. One audio recording of clean speech and a set of 1–3 recordings of noise samples of different types were selected. A random value of the signal-to-noise ratio of the noisy sample was chosen uniformly from a range of −5 to 40 dB. The amplitude of the sum of noise signals was scaled so that the total additive noise component gave the chosen signal-to-noise value. No multiplicative distortion was applied to the noisy signal. The noise records were scaled so that the signal-to-noise ratio of the total of the clean and noise components corresponded to a value randomly chosen from a uniform distribution in the range of 0 dB to 40 dB. There was no artificial reverberation. The set of test samples consisted of 150 noisy signals with signal-to-noise ratios ranging from 0 to 20 dB.

3.2 Implementation Details

The Python language and PyTorch deep learning library were used to implement the proposed model in Python. Two Nvidia GeForce 1080 Ti GPUs were used for the training and quality evaluation of the proposed neural network model. An Adam optimizer with a learning rate of 10^{-4} was used.

The training samples were cropped to a 3 s duration. In calculating the short-time Fourier transform, the audio signal was split into 32 ms frames with 50% overlap. A Hann function was used for windowing the frames.

The delay value τ was 2, as in other real-time speech enhancement models [2, 3, 5, 7], corresponding to a time delay of 32 ms at the selected frame durations. As in the previous work [3], the widths of the vector transformations S_G and S_X were 1 and 15 respectively. Repeating layers consisting of linear and S4 transformations (Figs. 2 and 3) had the following hyperparameter values. The output of the linear transform had a size of 512 and 384 applied for the full-band and sub-band blocks, respectively. The size of the hidden representation of the S4 transform was $N = 64$.

3.3 Performance Evaluation Metrics

The quality of the denoised speech signal is the main attribute of speech enhancement models. The most widely used measure of quality is calculated by the MOS technique [15]. The use of MOS requires the evaluation of audio recording set's quality by 20–60 expert listeners. A quality rating scale of 1 (unacceptable) to 5 (excellent) points is used. The average MOS of 4.0 or higher defines a good quality, where the reconstructed speech signal is usually indistinguishable from the original signal. The value, ranging from 3.5 to 4.0, guarantees speech quality sufficient for telephone communication.

In this paper, there was no possibility to conduct direct audial testing with the help of real listeners. Therefore, a set of standard automatically computed metrics was used to assess the quality of the enhanced speech. These included the Perceptual Evaluation of Speech Quality (PESQ) [16] in two versions (narrowband (NB-PESQ) and wideband (WB-PESQ) for signals with sampling rates of 8 kHz and 16 kHz, respectively), the scale-invariant signal-to-distortion ratio [17], and the short-time objective intelligibility (STOI) measure [18]. All these metrics are positively correlated with the MOS value.

3.4 Loss Function Choice

The choice of the most appropriate loss function was carried out by training the proposed model having n = 2 layers of S4-transformations on 100 epochs for each of the loss functions in Sect. 3. The results are presented in Table 1. It can be seen that the best result on all the metrics was shown by the model that was trained with the MSE loss calculated on masks.

Table 1. Performance comparison for the loss functions selection.

Loss function	STOI, %	SI-SDR, dB	NB-PESQ	WB-PESQ
MSE	**95.95**	**17.41**	**3.287**	**2.697**
-SI-SDR	94.82	16.93	3.107	2.516
L1-loss	95.64	16.98	3.271	2.667

3.5 Optimal Number of S4 Layers

Further, in case of the chosen MSE loss function, the choice of the optimal number of layers of S4-transformations was carried out. Versions of the model having n = 1, 2 and 3 similar Linear-S4 layers were considered. A further increase in the number of layers did not make sense, since the number of model parameters already at n = 4 would exceed the number of parameters of the model [2] (5.6×10^6 parameters), which was used as a baseline for comparison. Training was carried out for 100 epochs. The comparison results are presented in Table 2. An increase in the number of S4 layers in full-band and sub-band blocks leads to an increase in the quality for all the metrics. The model having 3 layers of S4 transformations in each block was chosen as the main one.

Table 2. Performance comparison for different numbers of S4 layers.

n	Model size, × 106	STOI, %	SI-SDR, dB	NB-PESQ	WB-PESQ
1	1.8	95.73	16.98	3.279	2.667
2	3.2	95.95	17.41	3.287	2.697
3	4.8	96.53	17.64	3.295	2.767

3.6 Comparison of the Proposed Model Having Alternatives

Performance metrics for the proposed model are shown in Table 3 along with comparisons to the top alternative models. The greater the metrics' values, the higher the model's quality. The lower limit of speech quality was represented by the values of the quality metrics that were calculated on the noisy signals in the top row (they were labelled as "Noisy" in Table 3). FullSubNet [2] and TS-LSTM [3] were two models in this comparison that were built using recurrent neural networks. As a result, they were able to process the signal quickly and in real time. A mixed recurrent-convolution architecture called DCCRN-E [5] was modified to enable handling the operations conducted on complicated values. Conv-TasNet [7] was a convolutional end-to-end compact network with a very little latency.

According to all the speech enhancement quality metrics, our S4-based model demonstrated superiority. At the same time, the size of this model, expressed by the number of parameters, is less than or comparable to most alternative approaches. It is especially important that, in comparison with the recurrent neural network methods [2, 3], a gain in the model size was demonstrated when using S4 transformations compared to LSTM layers, which, together with the use of the convolutional representation of the S4 transformation, in practice leads to a 1.2–1.5-time acceleration of the learning process.

Table 3. Performance comparison of the proposed and alternative approaches.

Model	Model size, $\times 10^6$	STOI, %	SI-SDR, dB	NB-PESQ	WB-PESQ
Noisy	–	91.52	9.071	2.454	1.582
FullSubNet	5.6	96.11	17.3	3.305	2.777
TS-LSTM	7.5	96.05	17.6	3.338	2.832
DCCRN-E	3.7	–	–	3.266	–
Conv-TasNet	5.1	–	–	–	2.730
S4-based	4.8	**96.53**	**17.64**	**3.295**	**2.767**

3.7 Discussion

Figures 4 and 5 show examples of the proposed method intended for improving the speech quality in comparison with the FullSubNet [2] recurrent network. In the case of low-frequency pollution (Fig. 4a), the model having S4 layers does not always completely remove the noise in the pauses (the time interval ranging from 1.6 s to 2.2 s in Figs. 4b and 4c). However, it restores the formant structure more accurately than it does the base model (the time interval of 2.2–2.7 s in Figs. 4b and 4c).

Non-stationary musical noises (Fig. 5) are quite well cleaned out by both models. One can see, however, that the proposed model does not accurately remove non-vocal sounds in pauses, leaving parasitic harmonics in the signal.

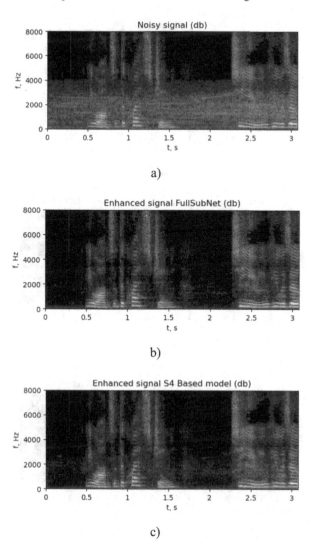

Fig. 4. An example demonstrating the low and mid-frequency noise (sawing sounds): a) noisy signal; b) result of the method from [2]; c) result of the proposed method.

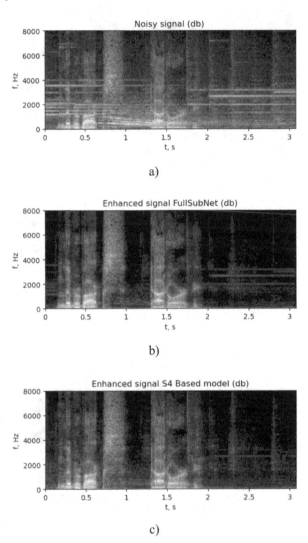

Fig. 5. An example involving the musical noise: a) noisy signal; b) result of method from [2]; c) result of the proposed method.

4   Conclusion

In this paper, a new method intended for speech enhancement based on the use of a two-stage neural network model having S4 transformations was proposed. It demonstrates a high quality of work compared to alternative approaches. Its main advantages are the relatively small size of the model, coupled with the ability to effectively use it in the real time regime. The proposed model can be applied in a wide range of real practical problems. It is promising for knowledge distillation usage among more complex neural network models used for a finding a more efficient solution to the problem of noise suppression.

Acknowledgements. This work was supported by the grant of the Russian Science Foundation, project no. 22-21-00199, https://rscf.ru/en/project/22-21-00199/.

References

1. Loizou, P.C.: Speech Enhancement: Theory and Practice. CRC Press, Boca Raton (2007)
2. Hao, X., Su, X., Horaud, R, Li, X.: FullSubNet: a full-band and sub-band fusion model for real-time single-channel speech enhancement. In: IEEE International Conference on Acoustics, Speech, and Signal Processing (ICASSP), Toronto, Canada, 6–11 June 2021, pp. 6633–6637 (2021). https://doi.org/10.1109/ICASSP39728.2021.9414177
3. Nasretdinov, R., Ilyashenko, I., Lependin, A.: Two-stage method of speech denoising by long short-term memory neural network. In: Jordan, V., Tarasov, I., Faerman, V. (eds.) HPCST 2021. CCIS, vol. 1526, pp. 86–97. Springer, Cham (2022). https://doi.org/10.1007/978-3-030-94141-3_8
4. Tan, K., Wang, D.: A convolutional recurrent neural network for real-time speech enhancement: In: Proceedings of the INTERSPEECH 2018, Hyderabad, India, 2–6 September 2018, pp. 3229–3233 (2018). https://doi.org/10.21437/Interspeech.2018-1405
5. Hu, Y., et al.: DCCRN: deep complex convolution recurrent network for phase-aware speech enhancement. In: Proceedings of INTERSPEECH 2020, Shanghai, China, 25–29 October 2020, pp. 2472–2476 (2020). https://doi.org/10.21437/Interspeech.2020-2537
6. Umut, I., Giri, R., Phansalkar, N., Valin, J.-M., Helwani, K., Krishnaswamy, A.: PoCoNet: better speech enhancement with frequency-positional embeddings, semi-supervised conversational data, and biased loss. In: Proceedings of INTERSPEECH 2020, Shanghai, 25–29 October 2020, pp. 2487–2491 (2020). https://doi.org/10.21437/Interspeech.2020-3027
7. Luo, Y., Mesgarani, N.: Conv-TasNet: surpassing ideal time-frequency magnitude masking for speech separation. IEEE/ACM Trans. Audio Speech Lang. Process. **27**(8), 1256–1266 (2019). https://doi.org/10.1109/TASLP.2019.2915167
8. Nasretdinov, R., Ilyashenko, I., Filin, J., Lependin, A.: Hierarchical encoder-decoder neural network with self-attention for single-channel speech denoising. In: Jordan, V., Tarasov, I., Shurina, E., Filimonov, N., Faerman, V. (eds.) HPCST 2022. CCIS, vol. 1733, pp. 3–14. Springer, Cham (2023). https://doi.org/10.1007/978-3-031-23744-7_1
9. Gu, A., Goel, K., Ré, Ch.: Efficiently modeling long sequences with structured state spaces. In: Proceedings of the Tenth International Conference on Learning Representations, ICLR 2022, Virtual Event, 25–29 April 2022 (2022)
10. Gu, A., et al.: Combining recurrent, convolutional, and continuous-time models with linear state-space layers. In: Advances in Neural Information Processing Systems 34 (NeurIPS 2021), Virtual Event, 6–14 December 2021, pp. 572–585 (2021)
11. Williamson, D.S., Wang, Y., Wang, D.: Complex ratio masking for monaural speech separation. IEEE/ACM Trans. Audio Speech Lang. Process. **24**(3), 483–492 (2016). https://doi.org/10.1109/TASLP.2015.2512042
12. Orvieto, A., et al.: Resurrecting recurrent neural networks for long sequences. arXiv (2023). https://doi.org/10.48550/arXiv.2303.06349
13. Lependin, A.A., Nasretdinov, R.S., Ilyashenko, I.D.: Speech enhancement method based on modified encoder-decoder pyramid transformer. Proc. Inst. Syst. Program. RAS (Proc. ISP RAS) **34**(4), 135–152 (2022). https://doi.org/10.15514/ISPRAS-2022-34(4)-10

14. Reddy, C.K.A., et al.: The INTERSPEECH 2020 deep noise suppression challenge: datasets, subjective testing framework, and challenge results. In: Proceedings of INTERSPEECH 2020, Shanghai, China, 25–29 October 2020, pp. 2492–2496 (2020). https://doi.org/10.21437/Int erspeech.2020-3038

15. International Telecommunication Union: ITU-T P.808 Subjective evaluation of speech quality with a crowdsourcing approach (2018)

16. Rix, A.W., Beerends, J.G., Hollier, M.P., Hekstra, A.P.: Perceptual evaluation of speech quality (PESQ) – a new method for speech quality assessment of telephone networks and codecs. In: IEEE International Conference on Acoustics, Speech, and Signal Processing. Proceedings, Salt Lake City, UT, USA, 7–11 May 2001, pp. 749–752 (2001). https://doi.org/10.1109/ICA SSP.2001.941023

17. Roux, J.L., Wisdom, S., Erdogan, H., Hershey, J.R.: SDR – half-baked or well done? In: IEEE International Conference on Acoustics, Speech and Signal Processing, Brighton, UK, 12–17 May 2019, pp. 626–630 (2019). https://doi.org/10.1109/ICASSP.2019.8683855

18. Taal, C.H., Hendriks, R.C., Heusdens, R., Jensen, J.: A short-time objective intelligibility measure for time-frequency weighted noisy speech. In: IEEE International Conference on Acoustics, Speech and Signal Processing, Dallas, TX, USA, 15–19 March 2010, pp. 4214–4217 (2010). https://doi.org/10.1109/ICASSP.2010.5495701

Designing a Graphics Accelerator with Heterogeneous Architecture

Ilya Tarasov$^{(\boxtimes)}$ (iD), Dmitry Mirzoyan, and Peter Sovietov (iD)

MIREA – Russian Technological University, Vernadsky Avenue 78, 119454 Moscow, Russia
`tarasov_i@mirea.ru`

Abstract. The article discusses the architecture of a graphics accelerator, based on a combination of general-purpose processor cores and pipeline accelerators for performing operations with matrices and transcendental operations. The article proposes a general architecture for GPUs of this type and suggests the main options for computing nodes designed to implement target group algorithms. In order to reduce technical and organizational risks, it is planned to simplify the hardware component of Very Large Scale Integrated Circuits (VLSI) and transfer the functions of managing calculations to embedded software, for which control processor cores have been introduced into VLSI. The VLSI project involves the development of a GPGPU-class computing accelerator, in which the ability to work with three-dimensional graphics is an additional feature. This allows you to take advantage of an architecture based on a large number of simple computational cores, using such VLSI in conjunction with a general-purpose processor.

Keywords: VLSI · Heterogeneous Architecture · Graphical Accelerator · GPU

1 Introduction

Currently, the relevance of creating an element base for high-performance computing is increasing. It is known that specialized computing devices are more efficient than general-purpose computers, but Very Large Scale Integrated Circuits (VLSI) with a limited scope of application have a smaller market and therefore turn out to be prohibitively expensive for small production runs. It is also important that, along with the development of technological standards of 3 nm and less technology node, the production of 28-7 nm standard VLSI and in some cases even larger standards (90–45), for applications with low performance requirements, but sensitive to development costs.

A widely known approach is based on the use of the GPU as a hardware accelerator working in combination with the CPU. Technologies such as Nvidia CUDA and OpenCL provide a layer of hardware abstraction which allows the use of a high-level C-like programming language. This expands the scope of use of CPU+GPU class computing systems. At the same time, GPUs can be considered as a type of architecture based on the use of a large number of simple computing cores grouped into clusters. This architecture can use combined connections at different levels of the hierarchy: tree, fat tree, ring, grid/mesh, etc.

© The Author(s), under exclusive license to Springer Nature Switzerland AG 2024
V. Jordan et al. (Eds.): HPCST 2023, CCIS 1986, pp. 29–40, 2024.
https://doi.org/10.1007/978-3-031-51057-1_3

Based on the significant complexity of designing VLSI, comparable in characteristics to solutions of the world's leading manufacturers (such as Nvidia and AMD), we can consider an alternative approach based primarily on the implementation of a computation accelerator, which additional function would be to work as a graphics coprocessor. It is worth noting that a number of GPU families are also focused on use as part of workstations with high performance in general-purpose tasks. Changing priorities will allow us to distance ourselves from performance assessments based on 3D graphics algorithms, since this is not the main purpose of such VLSI. At the same time, it becomes possible to implement architectural and circuit solutions that further enhance the capabilities of VLSI in target subclasses of tasks, where existing GPGPU VLSI are forced to also support data processing algorithms for 3D graphics. Studying the specifics of individual subclasses of algorithms and clarifying the current list of tasks that require hardware support is planned as part of the research work within the framework of VLSI design.

The following areas of application of specialized VLSI are considered:

- digital signal processing accelerators;
- software-defined radio;
- measuring instruments;
- medical equipment;
- accelerators for image processing;
- CCTV;
- industrial robots;
- machine learning;
- VR/AR.

2 Architecture of a Specialized Graphics Accelerator

According to Hennessey and Patterson [1], the dominant trend in performance improvement is Domain-Specific Architectures, DSA. At the same time, the need to place control components, and especially program memory, within the processor node increases the relative hardware costs for implementing one operation. Therefore, along with programmable computing nodes, non-programmable computing nodes designed to implement frequently used transformations can also be used as part of a specialized computing system. Non-programmable pipeline structures can be modified to be able to switch between separate operations at each stage, or to combine data movement along the pipeline with cyclic repetition of calculations at the same stage.

The combination of hardware and software methods for implementing calculations within the GPU was used both in Intel Larrabee projects [2] and in research projects based on the RISC-V core to implement general-purpose computing [3] or directly 3D graphics [4]. This gives reason to consider a similar approach using newly developed processor cores specialized for certain types of calculations as part of VLSI.

To design components that perform calculations as part of specialized VLSI chips, the following system-level implementation options can be considered:

1. Making changes to the data processing path of a specialized VLSI with a wide command word in order to provide support for operations typical for crypto conversions.

2. Connecting a pipelined data processing path to the processor core with a wide command word as an auxiliary arithmetic-logical device.
3. Connecting a pipelined data processing path to the processor core or system bus with a wide command word as a stand-alone configurable device.

The listed options can be considered as candidate architectures with clarification of their characteristics at the system level.

At the computing device architecture level, the following options can be considered:

1. Programmable computing node (processor).
2. Non-programmable (configurable) pipeline path for processing streaming data.
3. Combination of conveyor paths and programmable nodes.

The considered architectures of computing nodes are shown in Fig. 1.

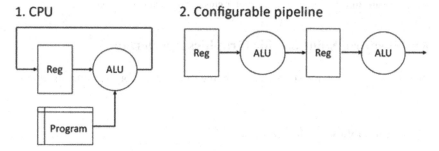

Fig. 1. Architectures of VLSI graphics accelerator computing nodes.

The architectures shown above correspond to mutually complementary approaches to organizing computing – CPU-based, i.e. distributed in execution time, and pipeline-based, distributed in the space of the VLSI chip. The current trend of using synchronous pipelined computers makes them preferable, but the functionality of a pipeline is determined by the order in which its stages are connected, while for a processor the order of calculations is determined by the program code and can be changed at runtime. At the same time, the hardware redundancy of the processor is determined by the need to add program memory in such a size that would ensure the execution of all algorithms of the target group.

In Fig. 2 the layout of a VLSI cluster that combines the functions of a graphics controller and a general-purpose computing accelerator is shown.

The architecture of the processor that implements the computing node is the subject of research. Compared to a general-purpose core (such as RISC-V), it is possible to further specialize the instruction set architecture while maintaining a simple microarchitecture. Reducing the redundancy of a single core will have a positive impact on the performance of VLSI chips that contain many such cores.

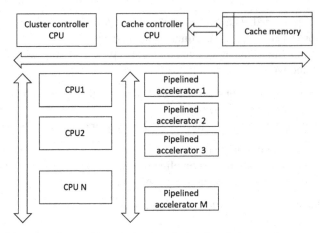

Fig. 2. Architecture of a graphics accelerator cluster aimed at general-purpose computing.

3 Architectural Solutions for Graphics Accelerator

The following architectural solutions are being considered for a promising graphics accelerator.

3.1 Programmable Task Distribution

GPU-specific computing tasks combine simple RISC-like operations and pipelined computing, as discussed in the previous section. A reduction in the complexity of the hardware component of the GPU control system can be achieved by transferring the task distribution functions entirely to the software component of the system. To perform this, a specialized task management processor is added to the VLSI cluster, which has access to the system bus and is controlled by both system drivers of the central processor and embedded software. The control processor runs cluster-local pipeline accelerators, if possible, or implements operations in software. To do this, it is necessary to provide access to the program memory of auxiliary processors based on dual-port memory.

3.2 Software Memory Management

The combination of on-chip static memory and external dynamic memory requires the implementation of a controller that, among other things, performs caching using certain algorithms. Depending on the scenarios for working with data, different caching algorithms may be optimal, which significantly complicates the design of the controller in the absence of a database of experimental data collected on implemented GPUs. Therefore, for a VLSI prototype project, software memory management is assumed with the allocation of address spaces and copying of data between memory of different types under the control of a dedicated processor.

3.3 Distribution of Tasks Depending on Data

If there are heterogeneous computational blocks, it becomes possible to assign a block not only in accordance with the type of task, but also, in some cases, depending on the specific values of the data being processed. For example, in a rotation matrix for angles that are multiples of 90 degrees, all coefficients are equal to 0, 1, or −1, which greatly simplifies multiplication by such matrices. Support for such operations is possible by introducing special flags that provide quick return of the result if one of the operands is 0, 1 or −1.

3.4 Redistribution of Resources within the Computing Cluster

Differences in resource requirements imposed by various algorithms of the target group under consideration necessitate the addition of memory and functional nodes, which will be redundant for a certain subclass of computing. Therefore, an approach based on placing a switched matrix of functional nodes, such as processor devices, memory and configurable pipelines, within one cluster is being considered. The ability to dynamically switch connections at medium and large levels of the hierarchy will allow memory to be redistributed between computing nodes, adapting VLSI to the requirements of the corresponding memory-intensive algorithms.

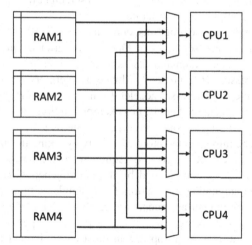

Fig. 3. An example of programmable distribution of memory blocks between processor cores.

In Fig. 3, N memory blocks are connected to M processor cores using full switches. In the mentioned scheme, it is possible both to distribute blocks in pairs across the corresponding processor cores, and to transfer all memory to one or more processor cores. This mode may be required to implement algorithms that require a large amount of memory. In this case, the overall performance of a group of processors will be reduced, since some of the cores are idle due to a lack of free memory blocks, but the ability to execute the algorithm remains.

In some cases, generalizing memory to form a larger block will not cause processors to turn off if the algorithm being executed is SIMD (Single Instruction, Multiple Data) class compliant.

3.5   Thread Management According to the SIMD Approach

Reducing the amount of required memory is possible by using the SIMD approach, in which the same program is used to control multiple compute nodes. This solution is suitable for a number of problems in three-dimensional graphics, digital signal processing (for example, multi-channel filtering) and mathematical modeling of processes using the finite element method.

3.6   Combining Raster and Cache Memory for General-Purpose Computing

For high resolution images (FullHD and 4K2K), storing pixels in a format that complies with the TrueColor standard requires a minimum of 24 bits per pixel, i.e. 48 Mbit for FullHD resolution and 192 Mbit for 4K2K. Additionally, we can use an alpha channel, as well as a hardware depth buffer with a resolution of at least 16 bits per pixel. This increases the total image buffer size to 96 or 256 Mbits of memory. If we consider the use of static memory, which with such a volume will occupy a large area, we should consider using this memory not only in the form of a screen buffer, but also as a general cache memory of the computation accelerator.

To represent general-purpose programs, one can use the SPIR-V language [5], which is used for intermediate representation of OpenCL programs. Since the language is designed as an intermediate language, its implementations are considered both for general-purpose processors [6] and for hardware accelerators designed on the basis of FPGAs. For example, [7] addresses accelerator integration, and other works explore the use of OpenCL for specialized areas. For example, digital filters [8], convolutional neural networks [9], and image motion prediction systems [10] use limited subsets of OpenCL, so they can be implemented more efficiently by taking this factor into account.

However, GPUs with OpenCL capability require an implementation of the full SPIR-V specification. This task is complicated by the heterogeneous nature of the operations supported in the language specification. For example, simple bitwise logical and arithmetic operations on integer arguments are easily implemented both as part of the arithmetic-logical unit of the processor core and as part of a pipeline, although pipelining such simple operations is inappropriate in most cases. Integer and floating point multiplication operations require pipelining, but can be implemented as hardware extensions to the processor core. Finally, transcendental functions (primarily trigonometric) implemented using the CORDIC algorithm require pipelined or cyclic implementation. Taking into account the features of subsets of SPIR-V commands, we can assume the joint use of the described approaches with the corresponding distribution of operations by type of implementing device, as shown in Fig. 4.

The integration of heterogeneous components within the processor subsystem was considered in [11]. The design route involves conducting pre-RTL modeling of the system to clarify its characteristics, followed by the implementation of parameterized

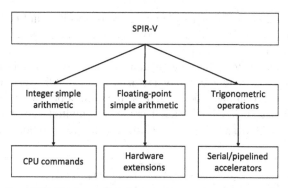

Fig. 4. Distributing SPIR-V intermediate language operations between cluster components.

components to perform individual operations. It appears promising to implement a configurable pipeline to perform operations based on the CORDIC algorithm, which form a large subset of SPIR-V instructions.

The general format of the command can be represented as follows:

$$<Dest> = <Operand1> \text{ op } <Operand2>$$

where *Dest* is the destination device (register) for storing the result of the operation; *Operand1* – first operand; *Op* – type of operation; *Operand2* – is the second operand.

For a command system, the concept of addressing is used, which reflects the number of registers described in the command code. Increasing addressability generally increases the capabilities of the tool software, but also increases the bit width of the command word, and therefore the amount of memory required to store the program. In this case, the absence of an indication of a particular type of resource means that it implicitly follows from the type of operation being performed or coincides with the specified resources. For example, in the instruction set of x86 processors, the destination register is the same as the first operand, in accumulator architectures the destination register and the first operand is always the accumulator, and in a stack architecture, the operands are always located on top of the data stack, and the result is also placed there.

In [12], a unified description of a computing node by four parameters (I, O, D, S) is considered, where I – number of instructions executed per clock cycle; O – number of operations determined by the instruction; D – number of operands (pairs of operands) related to operations; S – degree of conveyorization.

Based on this unified description, an ALU with a set of operations and combinations of operands is selected for the processor node, which:

1. Sufficient for implementing target group algorithms.
2. Optimal according to the selected criteria.

Optimality criteria include the number of clock cycles for executing algorithms, the area of VLSI (or the number of FPGA cells), the amount of program and data memory, and power consumption. The design route established for this project does not include early determination of the optimality criterion, since this limits the design space.

For a processor node, the register file model is not specified in SPIR-V and can be selected during the design process. This makes the number of registers and read/write ports parameters of the optimization process.

In Fig. 5 the transition to partial generalization of the resources of the pipeline stages is shown. If the pipeline generalizes only a register group, but the functional devices at individual stages have a similar structure and use the same subblocks, it may be possible to partially generalize such subblocks and implement multiplexers not between data paths, but between subblocks that are not identical in different versions of the data path.

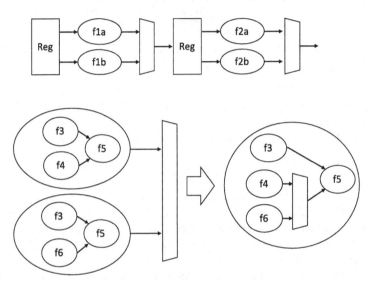

Fig. 5. Transformation of individual stages of a pipeline computer for the purpose of partial generalization of resources.

The implementation option of a conveyor with partial generalization of the resources of individual stages provides better component density, but at the same time complicates heat removal when using technological standards susceptible to the "dark silicon" effect. Therefore, the possibility of parameterized synthesis of pipeline stages should be maintained throughout the early stages of the project, right up to clarifying the characteristics of the topology library.

As part of preliminary research, the characteristics of a conveyor at stage 32, combining the performance of two types of operations, were assessed. The original RTL description of the module uses only one LUT layer using switching based on additional resources of logical cells - F7MUX, F8MUX. The pipeline is synthesized using 2247 LUTs and 2821 FFs based on the Xilinx Kria module. The trace results for a single pipeline are shown in Fig. 6.

An analysis of the placement of a pipeline with a set clock period of 1.5 ns shows that even in the absence of area constraints for the Xilinx Kria FPGA, the specified frequency is achieved due to the dense layout of the pipeline stages. An additional positive effect is the ability to shift the phase of the clock signal for individual registers (time borrow) for

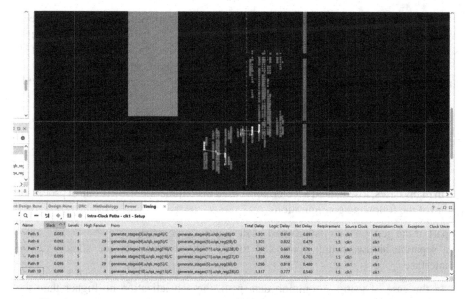

Fig. 6. Results of tracing a switched pipeline project without using area constraints

FPGAs made using 16 nm FinFET technology. This provides balancing of delays within the pipeline, implemented by CAD without direct involvement of the developer. Thus, pipeline structures are of interest as specialized devices for accelerating calculations, while providing compact placement of nodes and low latency due to local connections between individual stages of the pipeline.

4 Automation of Design of Specialized Nodes

Since the designed processor elements and pipeline computers are not standard, their development process is the subject to two opposing trends. On one hand, the improvement in functionality is determined by the complication of the ALU, an increase in the amount of processor resources and the complication of functions at individual stages of the pipeline. On the other hand, these actions lead to increased signal delays, die area, and power consumption. In the early stages of design, it is too difficult to create an accurate model of the target tasks, so priority should be given to the development of a VLSI system model that would allow the performance to be assessed for a certain combination of specified component parameters.

High performance of a hardware accelerator can be achieved by specializing its structure for those narrow classes of tasks where computational operations predominate over control operations and access to general data. At the same time, various forms of static parallelism are implemented in hardware form:

- task parallelism;
- data parallelism;
- pipelining.

High energy efficiency of specialized accelerators is achieved, in many cases, due to the irregularity of their structure, reflecting the specifics of a narrow class of problems, as well as through the use of local, direct connections between computing elements [13].

In this regard, first of all, specialization of the accelerator data path is of interest. It can be implemented using code analysis of target algorithms at various levels [14, 15]:

- an expression consisting of a small number of operations;
- linear section;
- cycle nest;
- hammock or function;
- call graph.

In many cases, a dedicated accelerator is used in conjunction with a control processor in one of the following configurations:

- part of the general data path as part of the control processor;
- a separate hardware unit connected to the control processor via some external interface.

In the case of using a common data path, the specialization no higher than the linear section level is the most preferable, since in this scheme a specialized accelerator competes with the control processor for shared resources, such as command fields, register file, memory.

Achieving the highest performance should be expected with hardware acceleration of calculations of complex software structures that include loop nests. Moreover, if the accelerator is implemented as a separate hardware unit, then the control processor is free to perform other tasks in parallel with specialized calculations.

To automate individual design tasks, a specialized CAD system is being developed, intended for the subclass of systems described in this article. The system uses descriptions of target algorithms in a high-level language [16] to analyze their features and distribute tasks between processor devices and pipelines. In addition to analyzing text representations, graphs are also currently used for this purpose [17], however, this method seems more labor-intensive if the volume of analyzed algorithms increases. Text analyzers are also used for this purpose [18–20].

The main tasks of CAD are:

- formation of a structural description of the upper level of VLSI;
- setting the parameters of VLSI components and formally checking their admissibility;
- integration with the compiler.

The technical specifications for CAD development are clarified as information is received about VLSI architectures, component parameters, assembly scenarios at the top level of description and other elements of the project, and design work that makes up the development process.

For the development of CAD, in accordance with the identified trends, a modular architecture is assumed in combination with a common project database. CAD elements can include both software applications developed in high-level languages and scripting languages, CAD scripts in these languages, as well as software interfaces of third-party tools such as compilers and applications for modeling domain processes.

5 Conclusions

The materials presented in the article represent the results of preliminary studies of the GPU architecture, intended to work as a computation accelerator as part of high-performance computing systems.

Acknowledgement. This work is supported by the Ministry of Science and Education of RF (Project No. FSFZ-2022-0004).

References

1. Hennessy, J.L., Patterson, D.A.: Computer Architecture. 6th edn. A Quantitative Approach (The Morgan Kaufmann Series in Computer Architecture and Design) (2017)
2. Seiler, L., et al.: Larrabee: a many-core x86 architecture for visual computing. IEEE Micro **29**(1), 10–21 (2009)
3. Elsabbagh, F., et al.: Vortex: OpenCL Compatible RISC-V GPGPU (2020). https://doi.org/ 10.48550/arXiv.2002.12151. Accessed 27 Feb 2020
4. Tine, B., Elsabbagh, F., Yalamarthy, K., Kim, H.: Vortex: extending the RISC-V ISA for GPGPU and 3D-graphicsresearch. In: Proceedings of MICRO-54: 54th Annual IEEE/ACM International Symposium on Microarchitecture, October 2021, pp. 754–766 (2021). https:// doi.org/10.1145/3466752.3480128
5. Khronos Group: Khronos SPIR-V Registry (2021). https://registry.khronos.org/SPIR-V/. Accessed 10 Oct 2023
6. He, W., et al.: Streamline ahead-of-time SYCL CPU device implementation through bypassing SPIR-V. In: Proceedings of the 2023 International Workshop on OpenCL, April 2023, Article no. 28 (2023). https://doi.org/10.1145/3585341.3585381
7. Leppänen, T., Lotvonen, A., Mousouliotis, P., Multanen, J., Keramidas, G., Jääskeläinen, P.: Efficient OpenCL system integration of non-blocking FPGA accelerators. Microprocess. Microsyst. **97**, 104772 (2023). https://doi.org/10.1016/j.micpro.2023.104772
8. Firmansyah, I., Yamaguchi, Y.: Real-time FPGA implementation of FIR filter using OpenCL design. J. Signal Process. Syst. **94**, 1–13 (2022). https://doi.org/10.1007/s11265-021-01723-6
9. Wu, Y., Zhu, H., Zhang, L., Hou, B., Jiao, L.: Accelerating deep convolutional neural network inference based on OpenCL. In: Shi, Z., Jin, Y., Zhang, X. (eds.) ICIS 2022. IFIPAICT, vol. 659, pp. 98–108. Springer, Cham (2022). https://doi.org/10.1007/978-3-031-14903-0_11
10. de Castro, M., Osorio, R., Vilariño, D., Gonzalez-Escribano, A., Llanos, D.: Implementation of a motion estimation algorithm for Intel FPGAs using OpenCL. J. Supercomput. **79**(5), 1–23 (2023). https://doi.org/10.1007/s11227-023-05051-3
11. Tarasov, I.E., Potekhin, D.S., Platonova, O.V.: Prospects for the use of soft processors in systems on a chip based on programmable logic integrated circuits. Russ. Technol. J. **10**(3), 24–33 (2022). https://doi.org/10.32362/2500-316X-2022-10-3-24-33
12. Sima, D., Fountain, T., Kacsuk, P.: Advanced Computer Architectures: A Design Space Approach. Addison-Wesley (1997)
13. Trilla, D., Wellman, J.-D., Buyuktosunoglu, A., Bose, P.: Novia: a framework for discovering non-conventional inline accelerators. In: Proceedings of 54th Annual IEEE/ACM International Symposium on Microarchitecture, October 2021, pp. 507–521 (2021). https://doi.org/ 10.1145/3466752.3480094

14. Zacharopoulos, G., Ferretti, L., Ansaloni, G., Di Guglielmo, G., Carloni, L., Pozzi, L.: Compiler-assisted selection of hardware acceleration candidates from application source code. In: Proceedings of 2019 IEEE 37th International Conference on Computer Design (ICCD), pp. 129–137. IEEE (2019)
15. Brumar, I., Zacharopoulos, G., Yao, Y., Rama, S., Wei, G.-Y., Brooks, D.: Early DSE and automatic generation of coarse-grained merged accelerators. ACM Trans. Embed. Comput. Syst. **22**(2), 32 (2021). https://doi.org/10.1145/3546070
16. Dave, S., Shrivastava, A.: Design space description language for automated and comprehensive exploration of next-gen hardware accelerators. In: Proceedings of Workshop on Languages, Tools, and Techniques for Accelerator Design (LATTE 2022) (2022)
17. Ferretti, L., Cini, A., Zacharopoulos, G., Alippi, C., Pozzi, L.: Graph neural networks for high-level synthesis design space exploration. ACM Trans. Des. Autom. Electron. Syst. **28**(2), 25 (2022). https://doi.org/10.1145/3570925
18. Agostini, N.B., et al.: An MLIR-based compiler flow for system-level design and hardware acceleration. In: Proceedings of the 41st IEEE/ACM International Conference on Computer-Aided Design, pp. 1–9 (2022). https://doi.org/10.1145/3508352.3549424
19. Venkataramani, G., Budiu, M., Chelcea, T., Goldstein, S.C.: C to asynchronous dataflow circuits: an end-to-end toolflow. Carnegie Mellon University, J. Contrib. (2018). https://doi.org/10.1184/R1/6603986.v1
20. Jordan, H., Scholz, B., Subotić, P.: Soufflé: on synthesis of program analyzers. In: Chaudhuri, S., Farzan, A. (eds.) CAV 2016. LNCS, vol. 9780, pp. 422–430. Springer, Cham (2016). https://doi.org/10.1007/978-3-319-41540-6_23

Spectrophotometer for Field Studies

Aleksandr Kalachev$^{(\boxtimes)}$ (ID), Vladimir Pashnev (ID), and Yuriy Matyuschenko (ID)

AltSU – Altai State University, Lenin Ave. 61, 656049 Barnaul, Russia
kalachev@phys.asu.ru

Abstract. A spectrophotometer for monitoring the condition of plants and vegetation covers has been developed. The device allows to determine the spectral composition of incident and reflected radiation and fluorescence spectra of plants, to calculate vegetation indices for numerical estimation of surface character or leaf condition. The spectrophotometer is realized as a BLE peripheral device. The structural diagram and description of individual components of the device are presented. Block diagrams of programs for working with sensors and interaction with the user are given. The instrument is designed as a compact portable device intended for operation in field conditions, has low power consumption and ergonomic interface. The spectrophotometer will find practical application in conducting research in the field of photosynthesis and plant biology, implementation of environmental monitoring activities, as well as in various branches of agriculture.

Keywords: Spectrophotometer · Vegetation Cover · Spectral Characteristics · Sensors · Wireless Microcontroller

1 Introduction

The most rational of "non-destructive" remote methods for plant condition monitoring are optical methods: by reflection or absorption coefficients of radiation and by fluorescence (in some cases by thermal radiation - the latter method is not widespread due to expensive equipment and specificity of its application, has no methods and is difficult to implement).

Effective non-invasive monitoring of plant condition is possible by tracking spectra:

- reflected radiation;
- incident radiation;
- fluorescence.

The aim of the work is to develop a compact spectrophotometer for field studies. The key requirements to the device are low power consumption, ergonomic interface.

Modern element base allows to form developed networks of microclimate monitoring of both open fields and greenhouse complexes. Moreover, recently appeared integral sensors allow to measure spectral composition of optical radiation. The size of the sensors allows them to be placed almost anywhere, including above/under plant leaves or on mobile platforms.

© The Author(s), under exclusive license to Springer Nature Switzerland AG 2024
V. Jordan et al. (Eds.): HPCST 2023, CCIS 1986, pp. 41–52, 2024.
https://doi.org/10.1007/978-3-031-51057-1_4

2 Proposed Approach

2.1 AS7341 Integral Spectrometer

The Ams AS7341 is an 11-channel spectrometer that enables new consumer, commercial and laboratory applications including spectral identification, reflection and absorption for color matching, liquid or reagent analysis, passive ambient light measurement and color calibration [4, 5]. The spectral response is defined by individual channels covering approximately 350 to 1000 nm, with 8 channels centered in the visible spectrum (VIS), plus one near-infrared (NIR) and broadband channel. The NIR channel, in combination with the other visible spectrum channels, can provide ambient light information. Additionally, the sensor has a specialized buffered channel for detecting light flicker down to a frequency of 2 kHz.

The optical interference filters of the AS7341 photodetectors are realized using nano-optical deposition technology. Figure 1 shows the center wavelengths of the AS7341 spectral channels.

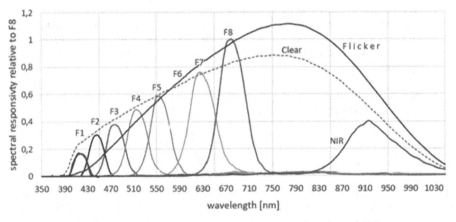

Fig. 1. Spectral characteristics of AS7341 measuring channels [4]:

F1: 415 ± 26 nm;	F6: 590 ± 40 nm;
F2: 445 ± 30 nm;	F7: 630 ± 50 nm;
F3: 480 ± 36 nm;	F8: 680 ± 52 nm;
F4: 515 ± 39 nm;	NIR (Near IR): 910 ± ~ 100–120 nm
F5: 555 ± 39 nm;	(not standardized in the documentation)

Figure 2 shows a structural diagram of the AS7341.

AS7341 compares favorably to similar spectrum sensors AS7262/AS7263 [1–3] with wider spectrum coverage, significantly lower current consumption (300 μA vs. 50 mA), requires less external elements. All this makes it logical to choose this device as a basis for spectrum sensors with autonomous (battery or accumulator) power supply. Some gap

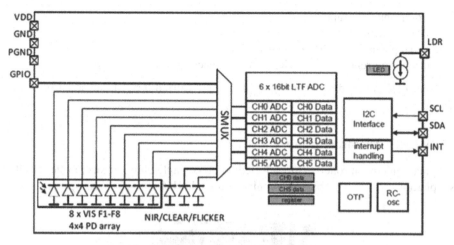

Fig. 2. Structural diagram of AS7341 [4].

in spectral channels in the red region of the spectrum (wavelengths 750 - 870 nm) can be attributed to the disadvantages. Figure 3 shows the appearance and structural diagram of the commercially available module Waveshare AS7341 [5].

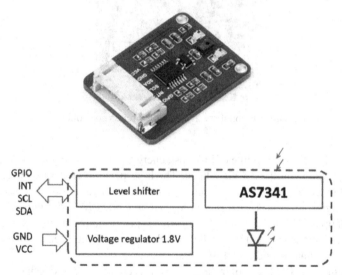

Fig. 3. External view and structural diagram of the Waveshare AS7341 module [5].

AS7341 Key Specifications:

– supply voltage 1.8 V;
– maximum current consumption 300 µA;
– sleep mode consumption less than 5µA;

- customizable interrupt sources;
- built-in 6-channel 16-bit ADC;
- external photodiode (GPIO) pin available;
- controlled current source (LDR) pin for connecting an LED.

Device control and access to spectral data arerealized through the serial interface I2C.

2.2 Wireless Microcontroller CC2640

As a main microcontroller's chip selected wireless single-chip microcontroller family SimpleLink – CC2640 as part of a miniature module HM-19 [6, 7] (Fig. 4).

Fig. 4. HM-19 module with CC2640 wireless controller [6, 7].

SimpleLink-CC264* is a line of low-cost, energy-efficient wireless controllers optimized for operations in the 2.4 GHz band. The high-performance transceiver is driven by a dedicated Cortex-M0 processor core executing low-level protocols flashed into its ROM. The higher-level protocols run on a separate Cortex-M3 core with a base clock frequency of up to 48 MHz. Sensor interrogation is performed by an independent micropowered hexadecimal-bit controller (the so-called Sensor Controller), which is capable of handling both analog and digital sensors.

At the same time CC264* controllers have almost unique technical characteristics in terms of performance, power consumption and radio channel parameters. An additional plus is the availability of CodeComposterStudio development environment and tool environment for this family of microcontrollers (SDK), TI-RTOS real-time operating system and peripheral device drivers. TI-RTOS allows application tasks to run in a multitasking environment and automatically manages the controller's power consumption modes.

In the current available versions of SDK CC2640 supports BLE protocol version 4.2, which will be used for device control and user interface organization.

2.3   Structure of the Spectrophotometer

The structural diagram of the spectrophotometer is shown in Fig. 5.

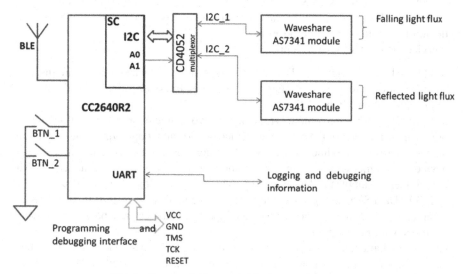

Fig. 5. Structural diagram of the spectrophotometer.

It consists of two Waveshare AS7341 modules for measuring incident and reflected fluxes. The upper window for incident flux is additionally covered with a frosted diffusing glass. Controlled LEDs of the modules allow to use active illumination for measuring the reflection coefficient of surfaces or for initialization of fluorescence.

The sensors operate via I2C interface, and since there is no possibility to change the AS7341 address on the I2C bus, hardware multiplexing of I2C lines is implemented with the help of analog two-channel multiplexer CD4052 series.

All sensor operation is organized through the SensorController (SC) sensor controller kernel. Several tasks are defined on the SC:

– independent change of spectrum sensor settings;
– running and obtaining measurement results from spectrum sensors;
– polling of temperature and humidity sensors.

2.4   Software Structure

Initially, the tasks are not active and are started by the CC2640 central core application depending on user actions. In Fig. 6 there are enlarged block diagrams of program operation for the CC2640 main core and for the sensor controller core. The work of the embedded application within the TI-RTOS framework together with the running BLE stack is organized according to the principle of event processing.

Two large groups of events can be distinguished: events generated by the protocol stack operation and events generated by user actions and tasks executed on the SC sensor controller.

The spectrophotometer is implemented as a BLE peripheral device and provides the device connected to it with a number of services and characteristics. The services have their own identifiers (UUIDs) from the range F000XXXXXX-0451–4000-B000–000000000000, where - XXXX is the 16-bit part of the service identifier and the rest is the 128-bit UUIDs namespace. In this case, the Texas Instruments namespace is used by default for demonstration purposes [8].

Service identifiers (XXXX - parts):

- 0x1110 - LED Service - LED and AS7341 settings management - contains two characteristics with register settings, one for each of the sensors, including the current of the external backlight LEDs;
- 0x1120 - Button Service - the service contains two characteristics with notification to organize the reaction of the connected device (smartphone/laptop/pressing the button on the measuring module; in case of changing the characteristic by the external device - starts measurements in one or simultaneously two spectral channels, at that the number of samples is determined by the new value of the characteristic;
- 0x1130 - Data Service - data exchange service - contains the results of current measurements, two characteristics with data of spectral channels, it is possible to enable the mode with notification.
- Change of characteristic values from outside causes the BLE-stack event "characteristic changed", which is placed in the event queue and processed in the main loop (Fig. 7).

Different LED service characteristics allow each of the spectral sensors to be independently configured. Changing them triggers a corresponding event and changes the AS7341 settings. Button events can be used to trigger measurements with default or current settings only.

A write to the Button service characteristics initiates the start of a series of several measurements in spectral channels. Reading of measurement results is asynchronous, as measurement tasks are completed on the SC core (Fig. 8).

Interaction with the user is realized by means of two buttons that allow to start measurements with preset parameters. Application on the host system (smartphone/laptop), implements a graphical user interface for setting the device operation mode and starting measurements.

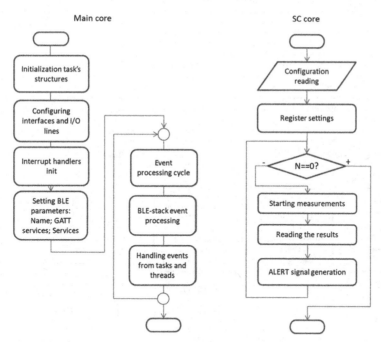

Fig. 6. Block diagrams of program operation for the main core of CC2640 and for the sensor controller core.

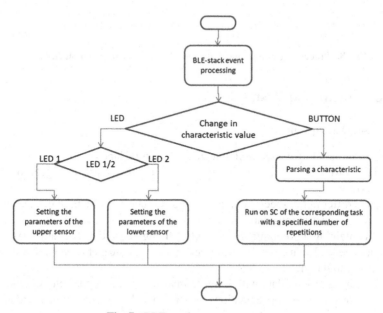

Fig. 7. BLE-stack event processing.

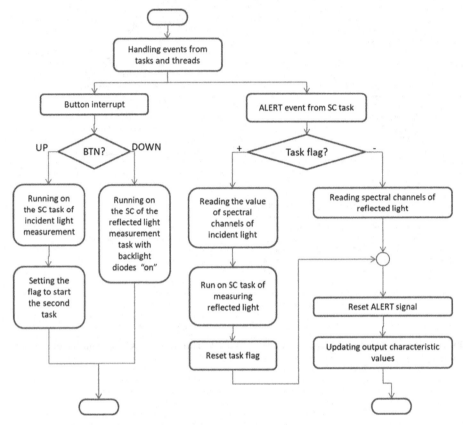

Fig. 8. Processing of button events and "ALERT" event from SC kernel.

3 Experiments and Results

3.1 Dataset Description

An example of recording the spectral reflectance of natural landscapes is shown in Fig. 9. Measurements were carried out at different sites with characteristic vegetation within the same territory.

The plots differ in the species composition of vegetation, which, in turn, affects the character of the spectral reflectance.

A site containing a predominant amount of living vegetation is characterized by a more pronounced reflectance peak in the green region and a steeper increase in reflectance in the near infrared region.

By comparing ground data with satellite imagery data it is possible to make more accurate and detailed maps of vegetation distribution and character, to track and identify the causes of anomalous values, to assess the content and character of aerosols in the atmosphere.

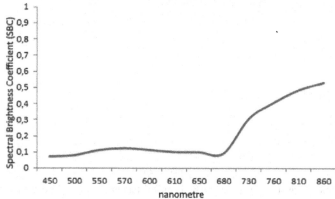

Fig. 9. Example of registration of spectral reflectance of natural landscapes in the Arctic region – moss and sedge vegetation with insignificant amount of cloudberries.

An example of recording the spectral reflectance of natural landscapes is shown in Fig. 10. Measurements were carried out at different sites with characteristic vegetation within the same territory.

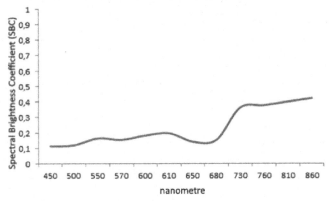

Fig. 10. Example of registration of spectral reflectance of natural landscapes in the Arctic region – Jagel-shrubby vegetation vegetation.

3.2 Discussion

Based on the available spectral channels, certain vegetation indices can be calculated using a spectrophotometer to provide a numerical assessment of surface character or leaf condition (Table 3) [9, 10].

The sensors used allow measuring the spectral composition of the radiation, and if necessary, they can also be used to measure the fluorescence of plants.

The spectrophotometer provides much more information about the light sources or the state of the vegetation under investigation. In addition, it can be used to measure the transmittance and absorption of light by samples from any external light source.

Table 1. Moss and sedge vegetation with insignificant amount of cloudberries

№	Types of plants	Types of plants. (Latin)	%	h↑, sm
1	Cloudberry	Rubus chamaemorus	15	15
2	Sphagnum	Sphagnum palustre	80	3
3	Pleuropod sedge	Carex chordorrhiza	60	20

Table 2. Moss and sedge vegetation with insignificant amount of cloudberries.

№	Ledum	Lédum palústre	%	h↑, sm
1	Cloudberry	Rubus chamaemorus	45	30
2	Yernik birch	Sphagnum palustre	80	3
3	Cladonia	Cladonia rangiferina	60	3
4	lingonberry	Vaccínium vítis-idaéa	3	15

Table 3. Performance of the proposed method in comparison with existing approaches.

Index	Formula
NDVI Normalized Difference Vegetation Index	$NDVI = (RNIR - RRED)/(RNIR + RRED)$
SR Simple Ratio	$SR = RNIR/RRED$
G Greenness Index	$G = R554/R677$
NPQI Normalized Phaeophytinization Index	$NPQI = (R415 - R435)/(R415 + R435)$
PRI Photochemical Reflectance Index	$PRI = (R531 - R570)/(R531 + R570)$
NPCI Normalized Pigment Chlorophyll Index	$NPCI = (R680 - R430)/(R680 + R430)$

RNIR – reflected signal in Near Ifrared light; RRED – reflected signal in Red light;
R X – denotes reflected signal with wavelength of X nm (so, R435 is a reflected signal on 435 nm)

4   Conclusion

A spectrophotometer for field research has been developed and tested.

An integrated approach has been applied to the registration of optical radiation fluxes - incident and reflected radiation are simultaneously recorded.

Integral sensors are used as primary sensors, which results in a minimum number of mechanical elements that increase the weight of the final product. The absence of mechanical filters and diffraction gratings increases vibration stability, reduces the weight of the product.

Target applications:

- photosynthesis research;
- plant biology research;
- plant screening, field work;
- environmental monitoring;
- agriculture and horticulture.

References

1. Kalachev, A.: In the footsteps of astrobotanics (in Russ.) (Po sledam astrobotanikov) (2018). https://www.rlocman.ru/review/article.html?di=514151. Accessed 10 Oct 2023
2. Sparkfun: AS7263-AS7262 6-Channel Visible Spectral_ID Device with Electronic Shutter and Smart Interface (2016). https://ams.com/documents/20143/36005/AS7262_DS000486_2-00.pdf/0031f605-5629-e030-73b2-f365fd36a43b. Accessed 10 Oct 2023
3. Sparkfun: AS7263–6-Channel NIR Spectral_ID Device with Electronic Shutter and Smart Interface (2016). https://cdn.sparkfun.com/assets/1/b/7/3/b/AS7263.pdf. Accessed 10 Oct 2023
4. Waveshare: AS7341: Datasheet (English). https://waveshare.xn--netw-nf7a/upload/f/f9/AS7341.pdf. Accessed 10 Oct 2023
5. Waveshare: AS7341 Spectral Color Sensor. https://waveshare.xn--comwiki-tb7c/AS7341_Spectral_Color_Sensor. Accessed 10 Oct 2023
6. WhizzBizz: HM-18/HM-19 CC2640R2 Bluetooth Module (2018). https://www.whizzbizz.com/p-httpd/multimedia/HM_18_HM_19_en_V1.pdf. Accessed 10 Oct 2023
7. Texas Instruments: CC2640 SimpleLink Bluetooth Smart Wireless MCU (2015). http://ebv news.xn--rudoc15-sb7c/cc2640.pdf, accessed on 10.10.2023
8. Kalachev, A.V., Lapin, M.V., Pelikhov, M.E.: Basics of working with Bluetooth Low Energy (in Russ.) (Osnovy raboty s tehnologiej Bluetooth Low Energy). Lan, St Petersburg (2022)
9. Dubinin, M.: Vegetation Indices (in Russ.) (Vegetatsionnye indeksy) (2006). https://gis-lab.info/qa/vi.html. Accessed 10 Oct 2023
10. Cherepanov, A.S.: Vegetation indices. Geomatics **28**(2), 98–102 (2011)

Comparative Study of Practical Implementation of Time Delay Estimation Methods on Single Board Computer

Vladimir Faerman[✉] ⓘ, Kirill Voevodin ⓘ, and Valeriy Avramchuk ⓘ

Tomsk State University of Control Systems and Radioelectronics, 40 Lenina Avenue, 634050 Tomsk, Russia
fva@fb.tusur.ru

Abstract. The article discusses the practical implementation of various methods for time delay estimation (TDE) on a Raspberry Pi single-board computer. The relevance of the research is due to the importance of the implementation of TDE methods in the tasks of object positioning and localization. The demand for real-time operation, as well as the requirement to use single-board computers as sensor nodes, imposes high demands on the efficiency of computing. The paper compares various time-domain and frequency-domain TDE methods, including those that utilize a limited set of spectral bins, applicable to problems of localization of acoustic signal sources. The paper considers various methods, their advantages, disadvantages, and computational features. In addition, we have carried out their comparative analysis as well as conducted experimental validation of theoretical estimates of the demands on computing resources. During a series of computational experiments carried out through specially developed software, the computing time and the memory usage are estimated. Based on empirical research on a single-board computer, the Raspberry Pi 4B, we reasonably advise certain methods to be employed in particular scenarios for localization of an acoustic source in space using the Raspberry Pi single boards.

Keywords: Time Delay Estimation · Raspberry Pi 4 · Fast Fourier Transform · Goertzel Algorithm · Computational Grids · Sliding Discrete Fourier Transform

1 Introduction

The significance of effective methods for time delay estimation (TDE) lies primarily in their wide practical application in local positioning systems [1, 2]. Schematically, the problem of local positioning is presented in Fig. 1 [1]. Depending on the specific circumstances within the scenario, positioning tasks can be considered as passive or active. In passive tasks, the position of the object that is the direct emitter of the signal is determined. In active tasks, the mobile object being positioned reflects a dedicated signal emitted by the locator [3]. It is also possible to reverse the composition of the system shown in Fig. 1, where the mobile object will be considered as the signal receiver. In turn, stationary nodes will become transmitters [4]. It should be noted that the reverse passive scenario is not significantly different from the regular passive one when considering solely the TDE problem. Time delay estimation methods are normally applyed to all those scenarios [1].

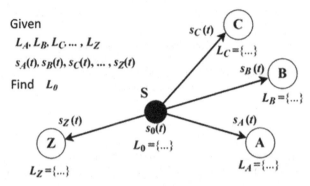

Fig. 1. Generalized TDE scheme (passive non-reverse scenario): S – mobile object (active emitter); A, B, C, ..., Z – array of stationary sensors; $s_0(t)$ – signal emitted by the source S; L_A, L_B, L_C, ..., L_Z – exact positions of the corresponding sensors; $s_A(t)$, $s_B(t)$, $s_C(t)$, ..., $s_Z(t)$ – signals received by corresponding sensors; L_0 – position of the source S.

The high demand for object positioning systems, in turn, is associated with the development of wireless technologies and advances in industrial automation. Recent development of the Internet of Things [5, 6] and the spread of smart devices [7, 8] have opened a great new field of application of positioning systems. The gradual introduction of such devices and systems into the consumer goods market makes it relevant to further reduce their cost. The principal factor of cheapening in this case is the reduction in the cost of hardware for smart devices or sensor nodes [6]. Systems on a chip (SoC) and SoC based single board computers are often considered as core computing units for positioning systems. The advantage of the latter is the developed peripherals and wide availability of hardware expansion modules, sufficient to solve most user tasks [9].

The main reason for choosing single-board computers is their high portability and self-sufficiency, as well as low power consumption and reduced cost compared to personal computers. However, you have to trade this for worse performance characteristics and lower volume of available RAM. This has to considered when implementing digital signal processing algorithms for practical applications.

In particular, a specific feature that must be taken into account when implementing local positioning techniques is the limited performance of the hardware platform, especially when operating in real time. This makes it relevant to study the practical methods of TDE and variants of their practical implementation in relation to various scenarios of application. The expected result is the reasonable choice of the TDE methods and their variants applicable to such scenarios as real time positioning of a mobile object, positioning of a signal source with *a priory* known spectral composition, and some others.

This article presents a theoretical and empirical comparison of different implementations of time-domain and frequency-domain methods of TDE in terms of computational performance and the amount of memory involved. In the experimental part of the work, a Raspberry Pi 4B single board computer was used as a hardware for tests.

2   Overview of TDE Methods

Time delay estimation is the most common technique applied to the local positioning problem. The essence of TDE is to measure a time lag between signals received by an array of spatially distributed sensors. In the basic passive scenario, the source of those signal is a positioning object.

For the sake of certainty, let us assume that only two sensors are used. In this case, the signals received by the sensors are described by the following expression [10]

$$
\begin{aligned}
s_a(t) &= K_a \cdot s_0(t - \tau_a) + n_a(t), \\
s_b(t) &= K_b \cdot s_0(t - \tau_b) + n_b(t),
\end{aligned}
\tag{1}
$$

where s_0 is the signal emitted by the object; K_a, K_b are the attenuation coefficients of signals along the propagation path; τ_a, τ_b are the delays between the received and the emitted signals; n_a, n_b are additive noises, s_a, s_b are received signals. So the required lag time is expressed as

$$
\tau_{ab} = (t - \tau_a) - (t - \tau_b) = \tau_b - \tau_a.
\tag{2}
$$

The value τ_{ab} is often referred to as the time difference of arrival (TDOA). Wide class of positioning methods is based on the analysis of set of TDOA estimations [1]. Depending on the practical setting, they can be used to determine the coordinates of an object by the method of multilateration in linear coordinates [11–13], in plane coordinates [14, 15], or in three-dimensions [5, 8, 16]. Such problems are quite typical for such fields of science and technology as local positioning and indoor navigation, wide area positioning of mobile subscribers in communication networks [5, 8], locating of pipeline leaks with leak noise correlators [11–13].

The applied methods to the aforementioned problem can be divided into two groups: time and frequency [17]. This segregation of methods is based on differences in the form of representation of signals directly in their analysis.

2.1   Time-Domain Methods

Time-domain TDE methods usually assume analyzing the correlogram [3]. The lag time can be obtained as an argument of the correlation function at which it reaches its maximum value:

$$
\tau_{ab} = \arg\max \left(R_{ab}(\tau) \right),
\tag{3}
$$

where arg max is the operator for obtaining the argument value at which function is maximized; $R_{ab}(\tau)$ is the correlation function of s_a and s_b.

The conventional way of calculating the correlation function of sampled signals $s_a(t_i)$ and $s_b(t_i)$ applies the convolution theorem [11].

$$
R_{ab}(\tau_j) = \mathrm{F}^{-1}\left(\mathrm{F}^*(s_a(t_i)) \times \mathrm{F}(s_b(t_i)) \right),
\tag{4}
$$

where F is operator of discrete Fourier transform (DFT) operator; F^{-1} is operator of inverse DFT (IDFT); * denotes unary operation of element-wise complex conjugation; × denotes binary operation of element-wise product.

Despite the prevalence of the approach based on (3) and (4), the practical implementation of time-domain methods may vary in a sensible way. The variants differ mainly by the algorithm that is used to perform DFT and IDFT operations.

2.2 Frequency-Domain Methods

Frequency-domain methods extract the lag time directly from the cross-spectrum of the analyzed signals. The discrete spectra $S_a (f_k)$ and $S_b (f_k)$ of the sampled signals $s_a (t_i)$ and $s_b (t_i)$ are complex valued, therefore we can consider them as following [18]

$$S_{a,b}(f_k) = F(s_{a,b}(t_i)), \quad S_{a,b}(f_k) = X_{a,b}(f_k) \times \Phi_{a,b}(f_k), \tag{5}$$

where $X_a (f_k)$, $X_b (f_k)$ are amplitude spectra; $\Phi_a (f_k)$, $\Phi_b (f_k)$ are phase spectra. The amplitude spectrum carries information about the energy properties of the signal. The phase spectrum carries information about the temporal features of the signal, in particular, the time shift is reflected in it.

The cross-spectrum, respectively, has the form

$$S_{ab}(f_k) = S_a^*(f_k) \times S_b(f_k),$$

$$S_{ab}(f_k) = X_{ab}(f_k) \times \Phi_{ab}(f_k),$$

$$X_{ab}(f_k) = X_a(f_k) \times X_b(f_k), \quad \Phi_{ab}(f_k) = \Phi_b(f_k) - \Phi_a(f_k). \tag{6}$$

The phase component of the cross-spectrum $\Phi_{ab} (f_k)$ is used to extract information about TDOA. The following formula can be used directly to estimate the lag [18]

$$\tau_{ab} = \sum_k \Theta_{ab}(f_k) \cdot f_k \left/ 2\pi \sum_k f_k^2, \right. \tag{7}$$

where $\Theta_{ab} (f_k) = U[\Phi_{ab} (f_k)]$ is the result of applying the unwrapping operator U to $\Phi_{ab} (f_k)$. [18]

Expression (7) is established based on the full spectral representation of the signals. Taking into account the equivalence of the information contained in the time and frequency representations of the signal, it is correct to consider (7) as an analogue of (3). The mathematical identity of these methods in terms of the potentially achievable accuracy is shown in [19]. However, it should be noted that in practical cases, they give different results. This is due to both the difference between real signals and model signals, and the inevitable differences in their computational simulation [20].

In [20], an alternative variant (7) is proposed, which allows one to use an arbitrary set of samples of the phase cross-spectrum. This makes it possible to apply the frequency method in situations where noise prevails at low frequencies. The use of an alternative formula also makes it possible to determine the time lag without computing the entire spectrum. This feature is discussed in detail in the following section.

2.3 Signal Processing in Practical TDE

In practical cases, the basic methods of TDE have low accuracy due to contamination of the source signal by additive noise on the side of the receiving sensors. Reduction in the negative impact of noise can be achieved by averaging estimates of the spectral characteristics of signals. Each spectral estimate is obtained by the short-time Fourier transform method [10]. In general, the time windows at the input of the transformation can overlap and have a shared subset of signal samples. The expressions for computing the correlation function for (3) and the phase cross-spectrum for (7) will respectively take the following form:

$$R_{ab}(\tau_j) = F^{-1}\left(\frac{1}{Q} \cdot \sum_q \left[F^*\left(s_a^{(q)}(t_i)\right) \times F\left(s_b^{(q)}(t_i)\right)\right]\right), \tag{8}$$

$$\Theta_{ab}(f_k) = U\left[\frac{1}{Q}\sum_q \left[\Phi\left(s_b^{(q)}(t_i)\right) - \Phi\left(s_a^{(q)}(t_i)\right)\right]\right],$$

$$\Theta_{ab}(f_k) = U\left[\arg\left(\frac{1}{Q}\sum_q \left[F^*\left(s_a^{(q)}(t_i)\right) \times F\left(s_b^{(q)}(t_i)\right)\right]\right)\right], \tag{9}$$

where Q is the total number of time windows at the input of DFT; arg is the operator that returns argument of a complex number; $s_a^{(q)}$, $s_b^{(q)}$ are subsets of signal samples belonging to the time window with index q.

In addition to averaging spectral estimates, frequency-weight functions of the form $\Psi_{ab}(f_k)$ are used to further reduce the influence of noise [21]. The values of the samples of the frequency-weight functions are positive and do not exceed unity. Values $\Psi_{ab}(f_k)$ close to unity indicate that at this frequency bin f_k, the signal overall prevails over noise throughout the entire observation period. Frequency-weight functions can be used with both time-domain [11, 21, 22] and frequency-domain methods [12, 23] of TDE. Despite the variety of such functions, averaged spectral estimates are always used to get them. From a computational standpoint, obtaining additional spectral estimates does not differ from the cross-spectrum estimate used in (8). For this reason, weighing in the frequency domain is not considered in the course of the further experimental study.

2.4 Variants of Fourier Transform Implementation

As shown above, both time and frequency methods of TDE require spectral transformations. In the practice of digital signal processing, DFT algorithms are used for this. Some of the most effective solutions are classified as fast Fourier transforms (FFT). Among the latter, the Coolie-Tukey algorithm is the most well-known and widespread [24].

Despite the significant computational advantages of FFT over the straight computation of the DFT, its use has a number of inconvenient features. Firstly, most FFT algorithms impose restrictions on the number of samples at the input of the transform. Secondly, those algorithms allow only the computation of the entire spectrum. This is redundant in cases where the informative signal is localized in several *a priori* known frequency bins. Thirdly, the obtaining of new data (a forward shift of the time window by several samples) requires a full-fledged application of the FFT to obtain a new spectral estimate. This complicates the use of FFT when operating in real time. The use of small windows is not always acceptable, since the size of the time window is associated with frequency resolution and noise tolerance. In contrast, the use of large windows, in combination with calling the transformation every time new data arrives, creates a large computational load.

Special DFT implementations were proposed for all the cases described above, where the FFT is limited in application. In particular, the chirp Z-transform (CZT) makes it possible to obtain an arbitrary number M of spectral bins using time windows composed of an arbitrary number N of ticks [25]. It should be noted that this limitation of the FFT is not essential for applying to TDE problem so we do not consider CZT in further study.

The Goertzel algorithm was initially proposed to compute individual frequency bins within the signal spectrum [26]. The use of this algorithm in conjunction with the frequency-domain methods of TDE allows one to obtain a time lag without computing the entire spectrum. This feature gives a computational advantage in some TDE scenarios, and therefore we will investigate it further.

A recursive sliding DFT algorithm can be used to obtain spectral estimates in real time [27]. The advantage of this algorithm is the ability to reevaluate the already available spectral characteristics based on newly received data. We will further investigate the performance of SDFT and the corresponding amount of used memory to determine its possibility of application to TDE problem.

It should be noted that by the moment numerous different recursive DFT algorithms have been developed and described [28, 29], which remained beyond the framework of this study. However, the potential of applying a few of them to TDE problem is discussed in conclusion.

3   Computational Study

To determine the operational capabilities of a sensor node based on a single-board computer, a series of computational experiments was carried out. During the experiments, an array of dummy data was processed with time-domain and frequency-domain TDE methods described in the previous section.

Further in this section we present estimates of the computational performance and memory usage benchmarks related to the most critical stages of the implementation of the considered methods. The discussion section showcases a comparison between the empirical outcomes derived from empirical investigations and theoretical estimations. Operation limits for the TDE device that are implied from the study are also could be found within the discussion.

3.1   Raspberry Pi 4B Hardware

Computational experiments feature a Raspberry Pi 4B single-board computer [30] with a HiFiBerry DAC + ADC Pro expansion board [31] shown in Fig. 2. The Broadcom

BCM 2711 SoC is the core processing unit of the computer. This SoC incorporates a quad-core general-purpose processor with the Cortex-A72 microarchitecture and a VC6 graphics core, along with some peripheral components. The HiFiBerry sound card was used exclusively in some segregated tests to verify the operability in real time for specific input data rates and particular preset of computational parameters. Therefore, its characteristics are not significant in the context of this study.

Fig. 2. Raspberry Pi 4B with HiFiBerry DAC + ADC Pro sound card attached on top.

3.2 Testing Software

For the purpose of this study, we have developed software for automated experimentation and statistical preprocessing of acquired data. We elected C++ as the main programming language, which was used to unify the program interfaces and implement wrap distinct computational functions.

Performance critical software components were implemented in low level in C. Our algorithmic implementation of the TDE methods largely corresponds to the description given in Sect. 2. To implement the FFT, we have used the current version of the FFTW library, that is in fact commonly considered as the branch standard. We implemented software components for SDF and Goertzel transform in a low level based on the algorithms described in [26] and [27] respectively.

We implemented a special class dedicated to acquisition of statistical data on computation time. Raw time benchmarks were gathered on calls of execution method of wrapper class. Each benchmark was reiterated 150 times. Then, the raw data underwent statistical processing. For each benchmark, we recorded the minimum and maximum execution times, the average time, as well as the standard deviation of time. Sample code for gathering and processing benchmarks is shown in Fig. 3.

Only dynamic allocation was taken into account when evaluating memory usage. This is due to the fact that a fair share of memory usage is associated with storing in buffers time series, complex spectra and precomputed constants for transforms. Due to

```
#ifdef FFT_GOERZEL_NUMBER
    std::cout << "Goerzel freq number benchmark" << std::endl << std::endl;
    for( auto N_ : N )
    {
        std::vector< double > data( 1 << N_ );
        for( auto& sample : data )
        {
            sample = static_cast<double>( std::rand() ) / RAND_MAX - 0.5;
        }
        for( auto diff_ : diff )
        {
            transform::cpu::forward::Goerzel transform( 1 << N_, GRZ_INDEX_FIRST, GRZ_INDEX_FIRST + diff_ );
            transform.SetReal( data );
            for( size_t i = 0; i < NUM_BENCH_ITERS; ++i )
            {
                auto begin = std::chrono::steady_clock::now();
                transform.Execute();
                auto finish = std::chrono::steady_clock::now();
                auto duration = std::chrono::duration_cast< std::chrono::microseconds >( finish - begin ).count();
                bench.UpdateRes( duration );
            }
            bench.CalcStats();
            std::cout << "N: " << N_ << "\t" << "diff: " << diff_ << "  \t\t"
                      << "min: " << bench.GetMin() << " \t" << "max: " << bench.GetMax() << " \t"
                      << "mid: " << bench.GetMiddle() << "\t" << "CKO: " << bench.GetCKO() << std::endl;
            bench.Reset();
        }
        std::cout << std::endl;
    }
    std::cout << std::endl;
#endif // FFT_GOERZEL_NUMBER
```

Fig. 3. Visual Studio screenshot that shows implementation of time measurements.

their size, these data arrays have to be stored in dynamic memory. Such an approach to the evaluation of memory usage is tolerant to distortion by memory, that is used on the stack and not directly related to the algorithms under study. The influence of the latter could not be excluded if the entire memory associated with the process was used as an estimate.

The *Valgrind* software was used to collect data on the allocated memory [32]. This tool is a specialized memory management service, debug utility system and profiler for software developers. Its functions include but not limited to the search for memory leaks, register attempts to accesses beyond the boundaries of allocated areas or use of uninitialized memory, and the investigation of other memory-related bugs.

3.3   Estimation of Computation Time

The variant of DFT implementation heavily influences the performance of a TDE method. This follows from (3) and (8), as well as (7) and (9), which is coherent with acquired experimental data. Any of these TDE methods requires at least $2 \cdot Q$ DFTs. This computational operation significantly prevails in (9). Other operations are mostly computationally simple: element-wise products of complex values, a unitary element-wise taking argument of complex numbers and element-wise multiplication by a scalar value. On the other hand, (8), in addition to similar element-wise operations, requires a single execution of the IDFT, which is computationally equivalent to an additional forward DFT.

For this reason, we further provide runtime estimates related only to the implementation variants of DFT. The execution time of the rest of the operations is not of comparable interest, because it has an auxiliary effect on the performance of the TDE methods, and also usually depends on the size of the time window N linearly.

The estimations of FFT execution time for various sizes N of the time window are presented in Table 1. Here and further, the following designations are used: T_{min} – minimum computation time; T_{max} – maximum computation time; T_{ave} – mean computation time; ΔT – standard deviation (half width) of computation time. Since many random factors can negatively affect the calculation time, we chose the minimum time as the most reliable estimate for the purpose of performance comparison. Key benchmarks for FFT are shown in Fig. 4.

When estimating the execution time of the Goertzel transform, we have varied both the number of samples in time windows and the number of calculated frequency bins. Since a theoretically predicted linear dependence of the execution time on the number of frequency bins presented in all experiments, in Table 2 we showed the computation time for a single frequency bin. Key benchmarks for Goertzel algorithm of DFT are shown in Fig. 5.

Similarly, when estimating the execution time of the SDFT, we varied the number of samples in time windows as well as the overlap rate between adjacent windows. Since we predictably found a linear dependence of the computation time on the number of newly introduced time samples in the previous time window, we elected to present the computation time for a single sample in Table 3. Key benchmarks for SDFT are shown in Fig. 6.

3.4 Estimation of Memory Usage

Estimates for the memory usage are given only for DFT variants, for similar reasons. However, the memory requirements depend on a TDE method to a greater extent than the performance. For instance, the use of frequency weighting functions requires storing in memory several additional spectral estimates (usually power spectra) as well as a set of frequency coefficients. So time-domain methods require the storage of whole spectra and the full set of frequency coefficients, while frequency-domain methods can rely on a limited set of frequency samples that require less memory to store.

Empirical estimates of the memory usage are presented in Fig. 7. The results of the study indicate the slight superiority of the Goertzel transform in this aspect. The actual advantage of the latter may be higher, given that the volume of required memory is dependent on the number of computed frequency bins (see Fig. 8). However, if we elect not to preserve inputs with FFT we can even make memory its usage lesser than Goertzel for full spectrum case.

Figure 8 clearly shows that the memory required for Goertzel transform is linearly dependent on the number of bins that have to be calculated. The constant term in the linear equation tends to become less significant with the size of the time window.

Table 1. Time to compute full spectrum with FFT.

N, samples	T_{min}, mcsec	T_{max}, mcsec	T_{ave}, mcsec	ΔT, mcsec
256	3	13	3.080	0.823
512	6	14	6.053	0.651
1024	13	19	13.040	0.488
2048	29	48	29.147	1.551
4096	76	129	77.853	6.912
8192	197	299	204.227	17.033
16384	428	639	444.313	39.560
32768	1006	1522	1046.770	82.995
65536	2902	3500	2969.608	76.057
131072	8715	11136	9101.190	426.446

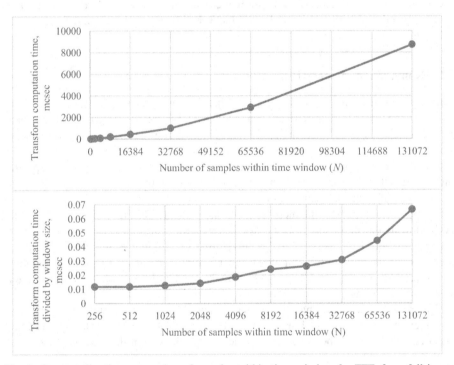

Fig. 4. Computation time vs number of samples within time window for FFT: for a full input window of N samples (on top); for a single input sample (on bottom).

Table 2. Time to compute one frequency bin with Goertzel transform.

N, samples	T_{min}, mcsec	T_{max}, mcsec	T_{ave}, mcsec	ΔT, mcsec
256	2	3	2.188	0.035
512	2	5	3.089	0.495
1024	5	6	5.753	0.032
2048	11	12	11.478	0.014
4096	22	24	22.756	0.110
8192	46	47	46.482	0.067
16384	93	95	93.306	0.111
32768	186	188	186.589	0.112
65536	373	374	373.163	0.030
131072	747	749	747.767	0.210

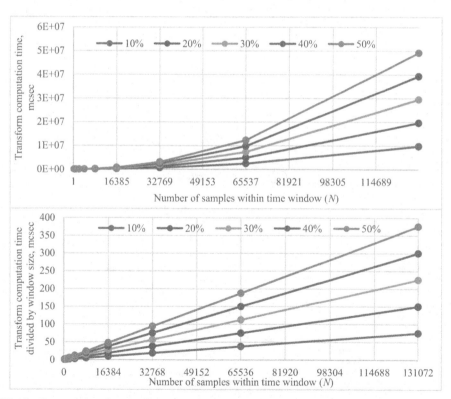

Fig. 5. Computation time vs number of samples within time window for Goertzel transform (various rates of computed frequency bins are indicated by the color of a curve): for a full input window of N samples (on top); for a single input sample (on bottom).

Table 3. Time to compute SDFT with almost overlapping time windows and precomputed spectrum for previous window (all samples but one are in both time windows).

N, samples	T_{min}, mcsec	T_{max}, mcsec	T_{ave}, mcsec	ΔT, mcsec
256	1	6	3.498	0.914
512	3	6	3.69	0.204
1024	7	9	7.277	0.11
2048	14	20	14.687	0.452
4096	29	30	29.551	0.039
8192	57	59	58.11	0.087
16384	115	117	115.945	0.117
32768	231	233	231.93	0.126
65536	487	491	488.799	0.692
131072	1085	1093	1088.615	1.269

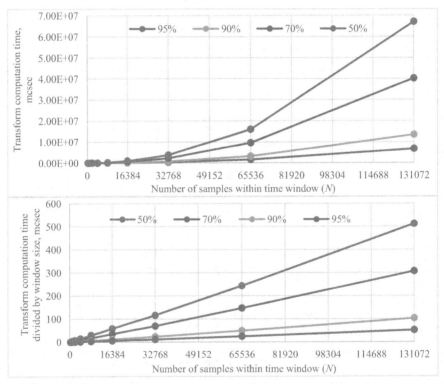

Fig. 6. Computation time vs number of samples within time window for SDFT (various rates of overlapping samples are indicated by the color of a curve): for a full input window of N samples (on top); for a single input sample (on bottom).

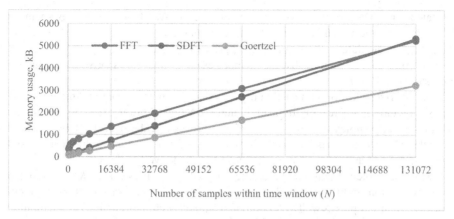

Fig. 7. Memory usage vs number of samples within time window for all considered DFT variants.

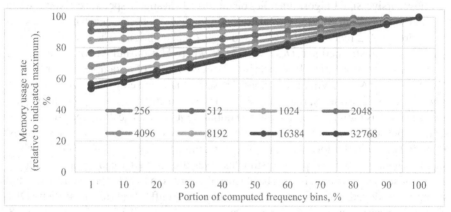

Fig. 8. Memory usage rate (compared to maximum value used for computation of full spectrum) vs rate of computed frequency bins for Goertzel transform (various windows size are indicated by the color of a lines).

4 Discussion

Our empirical results correspond well to theoretical estimations of complexity in regard to memory usage and computational operations required. Calculating the DFT for a real input time series using the FFT requires $(N/2)\cdot\log_2(N)$ complex multiplications and $N\cdot\log_2(N)$ complex additions. Each complex multiplication is composed of 4 real multiplications and 2 real additions. A complex addition is composed of 2 real additions. Thus, computing a spectrum via applying FFT to a time window of N samples requires $2N\cdot\log_2(N)$ real multiplications and $3N\cdot\log_2(N)$ real additions. The asymptotic computational complexity of the transform is $O(N) = N\cdot\log_2(N)$ and it is consistent with Fig. 4.

Each recursive FFT call requires splitting and reordering the interim results obtained at the current step of recursion. In the proposed implementation, a separate buffer was allocated for spectral data. However, it is possible to save about one third of memory by overwriting the initial sequence during computations. However, it is necessary to store the pre-calculated rotation multipliers prior to the transform, or performance will be compromised. Asymptotically, the memory usage of the FFT is $O(N) = N$, which is consisted with Fig. 7.

The Goertzel algorithm for calculating K frequency bins for an input time series of N samples requires $K \cdot 4N$ real multiplications and $K \cdot 5N$ real additions. Thus, its asymptotic computational complexity is $O(K,N) = K \cdot N$. Full scale real-valued DFT by Goertzel algorithm ($K = N/2 + 1$) is inefficient and requires $2(N^2 + 2N)$ real multiplications and $5(N^2/2 + N)$ real additions. These estimates correspond well to those curves presented in Fig. 5. The implementation of the Goertzel algorithm requires storing K precomputed complex rotation multipliers. At the same time, it is also necessary to store the input series of N real samples as well as K computed output spectral estimates. The asymptotic memory requirement of the Goertzel algorithm is $O(N,K) = N + K$, which is consistent with Fig. 8.

The recursive SDFT algorithm relies on the already available spectral estimates when recomputing spectrum with the arrival of new input data. Processing each new time sample requires $N + 2$ real multiplications and $N + 2$ real additions. The asymptotic complexity of the transformation is $O(M, N) = M \cdot N$, where M is the number of newly arrived non-processed time samples. Processing a full time window of N samples requires $N^2 + 2N$ additions and $N^2 + 2N$ multiplications. These estimates are consistent with the curves shown in Fig. 6. Like other variants, the sliding transform utilizes an array of rotation multipliers as well as buffers to store input and output sequences. The difference of SDFT is that the input series may not comprise a complete time window of all N samples in a first place. However, an additional internal buffer is required to store the N time samples which were used to obtain current spectral estimates. Even though SDFT can be called with any number of samples as input, it must be at least N samples in total before the first spectral estimate is produced. The requirement for an additional buffer leads to the fact that SDFT slightly underperforms in the aspect of memory requirements. That can be seen in Fig. 8. The asymptotic memory requirement of SDFT is $O(M,N) = M + N$.

Comparison of DFT implementation variants has shown that FFT is suitable in a wide range of scenarios, with few rare exceptions. Figure 9 shows the range of parameters of a computational problem in which Goertzel algorithms outperforms FFT. As far as a frequency-domain TDE method requires at least three frequency bins to draw a regression line, computational advantages can be achieved only by large time window sizes. This results in high frequency resolution. So to be practical in conjunction with a frequency-domain TDE method, SDFT requires an accurate *a priori* knowledge of the frequency localization of the signal as well as the absence of scattering during its propagation.

Figure 10 shows the range of parameters of a computational problem when SDFT has an advantage over FFT in execution time. One can infer from the figure that the use of the recursive algorithm is advisable only if an exceptionally high rate of spectrum recalculation is required. The sampling frequency is usually 44100 Hz if we assume a problem of positioning a mobile object via an acoustic channel. So in this case, the use of SDF will be practical only when the position of the object (along with spectral estimates) need to be updated at a rate exceeding 5000 Hz. Such a scenario seems not to be very realistic.

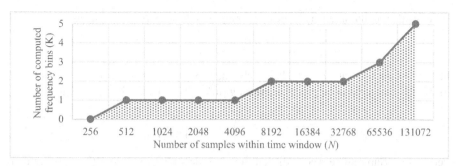

Fig. 9. Area in the domain of computational parameters when Goertzel outperforms FFT.

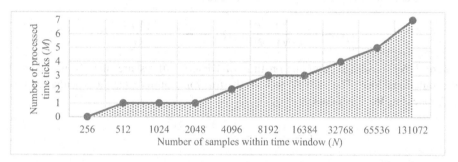

Fig. 10. Area in the domain of computational parameters when SDFT outperforms FFT.

A comparison of DFT implementation variants in the aspect of memory requirements showed that the Goertzel algorithm has a slight advantage. However, the amount of memory used is generally comparable for all variants, and this advantage appears to be of not high practical importance. If a critical limitation on memory is a case, then it is possible to save about one third of memory used by FFT just by giving up preserving input.

In the course of further discourse, we will assume that the functions of the sensor node are reduced to receiving continuously incoming signals, buffering them, processing them and output the results of their processing. Let us also assume that the intensity of data rate remains unaltered through all operating session, and the processing of the obtained results and their output are carried out asynchronously. The similar situation is described and modeled in [33].

Hence, real-time operating can be attributed to two parallel processes:

- processing of incoming data and refinement of spectral estimates (usually by coherent averaging of instantaneous spectra);
- and utilizing those spectra to measure time lags with TDE methods, use time lags to estimate object position solving multilateration problem.

The second process is not hardly synced with the first one and can be performed on demand or on a residual basis. On the contrary, acquiring and processing of incoming data should be done as soon as it arrives in order to avoid buffer overflow and data loss.

Let us denote the intensity of the incoming data flow as B and define as

$$B = f_d \cdot n,$$

where f_d – sampling rate; n – number of channels. So, the total number of time windows Q of size N that need to be processed during time period T_0 can be defined as

$$Q = T_0 \cdot \frac{B}{N} = T_0 \cdot \frac{f_d \cdot n}{N \cdot (1 - s)},$$

where s – overlap rate between adjacent time windows $(0 \leq s < 1)$. By supposing that the processing time of a window is predominantly determined by the DFT computation time, we estimate the total time T_Q takes to process all Q windows:

$$T_Q = Q \cdot \frac{T(N)}{k(N)} \cdot \frac{B}{N} = T_0 \cdot \frac{f_d \cdot n \cdot T(N)}{N \cdot (1 - s) \cdot k(N)},$$

where $T(N)$ – average computation time of DFT; k – the share of DFT in the total computational load. The ratio T_Q to T_0 is the fraction ρ of time that the machine spends on processing the input data stream:

$$\rho = \frac{T_Q}{T_0} = \frac{f_d \cdot n \cdot T(N)}{N \cdot (1 - s) \cdot k(N)}.$$

Let take that about two-thirds of computing resources need to be reserved in order to ensure timely and regular calculation and display of the object's position. Therefore, the number of channels that a computing device can serve can be roughly estimated as

$$n \leq \frac{N \cdot (1 - s) \cdot k(N)}{3 \cdot f_d \cdot T(N)}.$$

It is safe to assume that $k(N) \geq 0.66$ for any $N \geq 256$. We have previously determined the empirical values of $T(N)$ for the FFT and presented them in Table 1. A rough estimate using the formula above shows that the Raspberry Pi 4B is capable of acquiring and processing sound data from at least 16 channels at a frequency of 44100 Hz or from 4 channels at a frequency of 192000 Hz with an overlap rate of 75% in both cases. This potentially makes it possible to reevaluate the position of an object dozens of times per second, even using reasonably large size windows (for example, $N = 16384$). These qualitative estimates are confirmed empirically during test runs of Raspberry Pi 4 with a HiFiBerry module. More detailed and accurate benchmarks are planned for the future.

5 Conclusion

In this work, a comparative study of various implementations of TDE methods was made in respect of applicability on a single-board computer Raspberry Pi 4B. In the course of theoretical study of the issue and empirical research, it was established that DFT is the most computationally demanding operation, and it in a large extend determines the performance of the methods.

A comparison of various widely known DFT algorithms, in particular FFT, the Goertzel algorithm and the recursive SDFT algorithm showed significant advantages of FFT in performance and their equivalence in memory usage. Despite the fact that some areas of computational parameters in which the Goertzel algorithm and SDFT outperform the FFT, it is of little practical value.

The Raspberry Pi 4B single-board computer has sufficient computational capabilities to be used for positioning objects via an acoustic channel. The computer is able to process data streamed via 4 (or more) acoustical channels and to compute spectral estimates in a soft real time at a rate of at least 10 times per second. This way, FFT algorithms can be effectively utilized with time-domain and frequency-domain TDE methods.

In the future, other special DFT algorithms should be checked as well. In particular, the sliding Goertzel algorithm [34–36], which combines the features of both SDFT and the classical Goertzel algorithm, can probably make a competition to FFT in the scenario of tracking a mobile object in real-time.

References

1. So, H.C.: Source localization: algorithms and analysis. In: Position Location: Theory, Practice, and Advances, pp. 25–66 (2011). https://doi.org/10.1002/9781118104750.ch2
2. Gustafsson, F., Gunnarsson, F.: Positioning using time-difference of arrival measurements. In: ICASSP, IEEE International Conference on Acoustics, Speech and Signal Processing – Proceedings, pp. 553–556. IEEE (2003). https://doi.org/10.1109/icassp.2003.1201741
3. Chen, T.: Highlights of statistical signal and array processing. IEEE Signal Process. Mag. **15**, 21–64 (1998). https://doi.org/10.1109/79.708539
4. Hua, C., et al.: Multipath map method for TDOA based indoor reverse positioning system with improved Chan-Taylor algorithm. Sensors. **20**, 1–14 (2020). https://doi.org/10.3390/s20113223
5. Zhang, Y., Gao, K., Zhu, J.: A TDOA-based three-dimensional positioning method for IoT. Adv. Eng. Res. **149**, 775–779 (2018). https://doi.org/10.2991/mecae-18.2018.136
6. Tay, S.I., Lee, T.C., Hamid, N.Z.A., Ahmad, A.N.A.: An overview of industry 4.0: definition, components, and government initiatives. J. Adv. Res. Dyn. Control Syst. **10**, 1379–1387 (2018)
7. Hong, J.M., Kim, S.H., Kim, K.J., Kim, C.G.: Multi-cell based UWB indoor positioning system. In: Nguyen, N.T., Gaol, F.L., Hong, T.-P., Trawiński, B. (eds.) ACIIDS 2019. LNCS (LNAI), vol. 11432, pp. 543–554. Springer, Cham (2019). https://doi.org/10.1007/978-3-030-14802-7_47
8. Lee, K., Hwang, W., Ryu, H., Choi, H.J.: New TDOA-based three-dimensional positioning method for 3GPP LTE system. ETRI J. **39**, 264–274 (2017). https://doi.org/10.4218/etrij.17.0116.0554
9. Kurkovsky, S., Williams, C.: Raspberry Pi as a platform for the internet of things projects: Experiences and lessons. In: Annual Conference on Innovation and Technology in Computer Science Education, ITiCSE, pp. 64–69 (2017). https://doi.org/10.1145/3059009.3059028
10. Carter, G.C.: Coherence and time delay estimation. Proc. IEEE **75**, 236–255 (1987). https://doi.org/10.1109/PROC.1987.13723
11. Gao, Y., Brennan, M.J., Joseph, P.F.: A comparison of time delay estimators for the detection of leak noise signals in plastic water distribution pipes. J. Sound Vib. **292**, 552–570 (2006). https://doi.org/10.1016/j.jsv.2005.08.014

12. Faerman, V., Voevodin, K., Avramchuk, V.: Frequency-domain generalized phase transform method in pipeline leaks locating. In: Communications in Computer and Information Science, pp. 22–38. Springer, Heidelberg (2023). https://doi.org/10.1007/978-3-031-23744-7_3

13. Glentis, G.O., Angelopoulos, K.: Leakage detection using leak noise correlation techniques - Overview and implementation aspects. In: PCI '19: 23rd Pan-Hellenic Conference on Informatics, pp. 50–57. ACM, New York (2019)

14. Spencer, S.J.: The two-dimensional source location problem for time differences of arrival at minimal element monitoring arrays. J. Acoust. Soc. Am. **121**, 3579–3594 (2007). https://doi.org/10.1121/1.2734404

15. Dalskov, D., Olesen, S.K.: Locating acoustic sources with multilateration applied to stationary and moving sources (2014). http://www.es.aau.dk/sections/acoustics. Accessed 10 Oct 2023

16. Qu, J., Shi, H., Qiao, N., Wu, C., Su, C., Razi, A.: New three-dimensional positioning algorithm through integrating TDOA and Newton's method. EURASIP J. Wirel. Commun. Netw. **2020**, 77 (2020). https://doi.org/10.1186/s13638-020-01684-7

17. Björklund, S.: A Survey and Comparison of Time-Delay Estimation Methods in Linear Systems. Linköpings universitet, Linköping (2003)

18. Brennan, M.J., Gao, Y., Joseph, P.F.: On the relationship between time and frequency domain methods in time delay estimation for leak detection in water distribution pipes. J. Sound Vib. **304**, 213–223 (2007). https://doi.org/10.1016/j.jsv.2007.02.023

19. Zhao, Z., Zi-Qiang, H.: The generalized phase spectrum method for time delay estimation. In: ICASSP '84. IEEE International Conference on Acoustics, Speech, and Signal Processing, pp. 46.2.1–46.2.4 (1984)

20. Faerman, V., Avramchuk, V., Voevodin, K., Sidorov, I., Kostyuchenko, E.: Study of generalized phase spectrum time delay estimation method for source positioning in small room acoustic environment. Sensors. **22**, 965 (2022). https://doi.org/10.3390/s22030965

21. Knapp, C.H., Carter, G.C.: The generalized correlation method for estimation of time delay. IEEE Trans Acoust. ASSP **24**, 320–327 (1976)

22. Fuchs, H.V., Riehle, R.: Ten years of experience with leak detection by acoustic signal analysis. Appl. Acoust. **33**, 1–19 (1991). https://doi.org/10.1016/0003-682X(91)90062-J

23. Ma, Y., Gao, Y., Cui, X., Brennan, M.J., Almeida, F.C.L., Yang, J.: Adaptive phase transform method for pipeline leakage detection. Sensors (Switzerland) **19**, 310 (2019). https://doi.org/10.3390/s19020310

24. Frigo, M., Johnson, S.G.: FFTW: An adaptive software architecture for the FFT. In: Proceedings of the 1998 IEEE International Conference on Acoustics, Speech and Signal Processing, ICASSP 1998, pp. 1381–1384 (1998). https://doi.org/10.1109/ICASSP.1998.681704

25. Rabiner, L.R., Schafer, R.W., Rader, C.M.: The Chirp z-transform algorithm. IEEE Trans. Audio Electroacoust. AU **17**, 86–92 (1969). https://doi.org/10.7551/mitpress/5222.003.0015

26. Goertzel, G.: An algorithm for the evaluation of finite trigonometric series. Am. Math. Mon. **65**, 34–35 (1958)

27. Grado, L.L., Johnson, M.D., Netoff, T.I.: The sliding windowed infinite fourier transform: tips & tricks. IEEE Signal Process. Mag. **34**, 183–188 (2017). https://doi.org/10.1109/MSP.2017.2718039

28. Chicharo, J.F., Kilani, M.T.: A sliding Goertzel algorithm. Signal Process. **52**, 283–297 (1996)

29. Chauhan, A., Singh, K.M.: Recursive sliding DFT algorithms: a review. Digit Signal Process. **127**, 103560 (2022). https://doi.org/10.1016/j.dsp.2022.103560

30. Raspberry Foundation: Raspberry Pi 4 Model B (2021). https://datasheets.raspberrypi.com/rpi4/raspberry-pi-4-product-brief.pdf. Accessed 20 June 2023

31. Modul 9: Datasheet for HifiBerry DAC+ADC Pro (2020). https://www.hifiberry.com/docs/data-sheets/datasheet-dac-adc-pro/. Accessed 20 June 2023

32. Valgrind: Debugger and profiler valgrind-3.21.0. (2023). https://valgrind.org/downloads/. Accessed 10 Oct 2023

33. Faerman, V., Voevodin, K., Avramchuk, V.: Case of discrete-event simulation of the simple sensor node with CPN tools. In: International Siberian Conference on Control and Communications (SIBCON), pp. 1–9 (2022). https://doi.org/10.1109/SIBCON56144.2022.100 02956

34. Jacobsen, E., Lyons, R., Communications, H., Associates, B., Lyons, R.G.: Sliding spectrum analysis. In: Streamlining Digital Signal Processing: A Tricks of the Trade Guidebook, pp. 175–188. Willey-IEEE Press (2012). https://doi.org/10.1002/9781118316948.ch18

35. Sysel, P., Rajmic, P.: Goertzel algorithm generalized to non-integer multiples of fundamental frequency. EURASIP J Adv Signal Process. **2012**(1), 56 (2012). https://doi.org/10.1186/1687-6180-2012-56

36. Sridharan, K., Babu, B.C., Kannan, P.M., Krithika, V.: Modelling of sliding goertzel DFT (SGDFT) based phase detection system for grid synchronization under distorted grid conditions. Procedia Technol. **21**, 430–437 (2015). https://doi.org/10.1016/j.protcy.2015.10.065

Information Technologies and Computer Simulation of Physical Phenomena

Study of the Functional Characteristics of TiNi Coatings by the Computer-Aided Simulation Using Parallel Computing

Vladimir Jordan[1,2]([✉]) [iD] and Vitaly Blednov[2] [iD]

[1] Altai State University, Lenin Ave. 61, 656049 Barnaul, Russia
jordan@phys.asu.ru
[2] Khristianovich Institute of Theoretical and Applied Mechanics, SB RAS,
Institutskaya Str. 4/1, 630090 Novosibirsk, Russia

Abstract. The architecture and functionality of a program system that implements the paradigm of parallel execution of SIMD-tasks, in which a certain class of coatings with different parameter sets of particles sprayed onto the surface of technical products are simulated, is considered. Each SIMD-task uses a copy of the same computing module, which implements the computational foundations of a previously created software package for simulation the layered structure of a gas-thermal coating and calculating its functional characteristics. The program system allows you to run a series of computational experiments with the least amount of time due to the parallel running of SIMD-tasks for simulation a class of functional coatings, taking into account various spraying modes, and save electronic reports with illustrative research material in archival storage. The paper presents the results of simulation a class of coatings made of titanium nickelide (TiNi), taking into account various sets of "key physical parameters" (KPPs) of particles, characterizing different modes of coating spraying. In addition, based on the simulation results, an analysis was carried out of the variability of the coatings functional characteristics (porosity and adhesive strength of coatings, surface roughness of coatings) when varying the KPP-values in certain ranges. As a result of analyzing the simulation results of TiNi coatings and calculating their functional characteristics, optimal modes for the stable spraying of TiNi coatings on two types of substrates (on a Steel45 substrate and on a titanium substrate) were established.

Keywords: Program System · Computational Experiments · Parallel Computing · SIMD-Task · Simulation · Spraying · Gas-Thermal Coatings · Titanium Nickelide

1 Introduction

In the field of materials science, methods and technologies for gas-thermal spraying (GTS) of functional coatings (on technical products), which are characterized by a high degree of wear resistance, resistance to shock mechanical and thermal influences, etc., have now been quite effectively developed. Effective GTS-technologies are the following [1–3]: APS (atmospheric plasma spraying); D-gun spraying (detonation-gas spraying); HVOF-spraying (high velocity oxygen fuel spraying) and many other technologies that allow the formation of coatings from powders of metal, ceramic, or metal-ceramic particles (cermets).

Scientific publications and various monographs on this topic contain a fairly large volume of experimental data on the functional characteristics (physical properties) of gas-thermal coatings. However, so far such data have not been sufficiently systematized and, on their basis, a holistic understanding of the relationship between modes of spraying and the functional characteristics of the resulting coatings has not been formed. To solve this problem, the authors developed a program complex for simulation and designing the layered structure of gas-thermal coatings and calculating their functional characteristics [4, 5]. This program complex allows you to carry out sequentially one after another the computational experiments (CEs), corresponding to various sets of spraying modes parameters and initial data characterizing the physical properties of sprayed materials. However, the computational costs of performing computational experiments on modern computers are quite high (the time spent on one CE can reach several days). Therefore, the actual problem of increasing the computational efficiency of the software package was an attempt by the authors to develop a program system for parallel running of identical copies of a computing module that implements the computational foundations of the previously created software package for simulation the layered structure of gas-thermal coatings and calculating their functional characteristics. A series of CEs with a different sets of parameters corresponding to the spraying of a certain class of coatings is performed by the program system in parallel mode.

Turning to the history of the development of computer technology, in the first stages of its development with the use of computers in large scientific institutions, it became possible to develop programs that completely carry out the numerical solution of any particular version of a computational problem based on one set of data. When changing the data set, it was necessary to contact the computer center again with a request to complete the task. The results obtained were processed manually, then compiled into a report, and the missing intermediate values were supplemented by interpolation methods based on the calculated data.

Currently, the processor architecture in personal computers (PCs) is multi-core; processors and computers are combined into multiprocessor configurations in the form of large computing clusters and MPP systems. It became possible not only to increase the complexity of the problems being solved, but also to perform cycles of a large number of computational experiments (up to 1000 or more) using identical copies of one program with different sets of data (variable parameters – input data of a "typical" task). In addition, it is possible to summarize the calculation results into ready-made "electronic" reports. A cycle (series) of computational experiments (CEs), executed using a single program (for example, in Fortran, C/C++, etc.) with different sets of input data of the same structure and sizes, defines the so-called "class of similar SIMD-tasks".

This work aims to further improve and use for research in the field of materials science a program system, developed on the basis of the program complex previously created by the authors for simulation (computer-aided design) of gas-thermal coatings and calculating their functional characteristics (see Fig. 1). The program system makes it possible to automate the running of a series of CEs with the processing of their results and the preparation of electronic reports with the least amount of time due to the parallel execution (running) of the corresponding class of similar SIMD-tasks. Computational modules (CMs) in SIMD-tasks are represented by the same program code, which is the computational basis of the program complex for coating simulation (Fig. 1), i.e. all CMs are identical software copies.

2 Brief Description of the Program Complex for Coating Simulation

The composition of the program complex [6] designed for coating simulation is shown in Fig. 1.

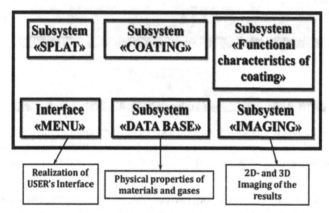

Fig. 1. Structure of the program complex for coating simulation and calculating their functional characteristics [6].

Let us consider the capabilities of the "SPLAT" subsystem and the "COATING" subsystem [4–6].

Subsystem "SPLAT" allows you to calculate the splat geometric characteristics (using the subsystem "DATA BASE"). The splats formed as a result of the collision of particles in the form of microdroplets of the melt with the base (substrate or pre-sprayed layer) with their subsequent spreading and solidification on the base (substrate). The calculation of the splat characteristics is carried out using a selected value set of "key physical parameters" (KPPs) of particles from the "DATA BASE" subsystem, which stores data on the physical properties of particle materials (see Fig. 1). In addition, subsystem "SPLAT" allows you to determine the critical ranges boundaries of KPP values, within which the splat "model" shape remains inseparable – this is important for adjusting the technological regime for spraying coatings formed as a result of fixing splats on the surface during the spraying process [4–6]. The calculation of splat characteristics is based on theoretical self-similar solutions for the process of forming one splat and which have undergone experimental verification. Self-similar solutions require the solution of a three-dimensional non-stationary boundary value problem with a free boundary for the Navier-Stokes equations. Solutions must be consistent with consistent with the equations of conjugate convective-conductive heat transfer and phase transformations in a spreading particle (in some cases, in the substrate) [4–6].

Subsystem "SPLAT" provides the opportunity for the subsystem "COATING" in the process of simulation the layered structure of the coating (at the stage of "laying splats" on the sprayed surface) to choose one of two model splat "morphologies" (Fig. 2): in the form of a disk and in the "smoothed" splat shape. In Fig. 2(b), the stepped profile of the smoothed splat shape is very noticeable, since of the three coordinates X, Y and Z, two coordinates X and Y take on discrete values with a given sampling step. The sampling step was deliberately chosen to be not small so that in Fig. 2(b) one could see the manifestation of the "stepping" of the splat profile [6].

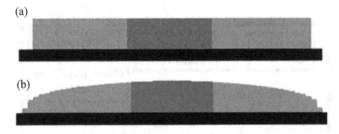

Fig. 2. Microsections of model splat morphologies (black color – substrate; dark gray shade – splat core; light gray shade – peripheral annular splat part): (a) – disk (cylindrical splat); (b) – smoothed splat, equivalent in volume to disk splat [6].

The shape of the splat (selected from two possible ones) at the stage of laying the splat on the surface is deformed in accordance with the topography (relief) of the local area of the surface on which the splat is formed during spreading and solidification (Fig. 3). Based on the relief of the surface area, the trajectory and angle of the melt flow into the recess is determined in accordance with certain formulas [5]. All empirical parameters in these formulas were estimated at the stage of their calibration using experimental data.

The "COATING" subsystem [4–6] models the process of creating a coating layered structure in accordance with the *algorithm for sequential stochastic laying of splats* (essentially, the Monte Carlo method) on the sprayed surface. Different areas of the sprayed surface may differ in relief (see Figs. 3(a), (b)). The dimensions and shape of the splat are determined by the "SPLAT" subsystem (Fig. 2), and in the laying process of the splat, the "COATING" subsystem (more precisely, the laying algorithm) deforms the shape of the splat according to the relief (topology) of the contact surface area (Fig. 3).

From Fig. 3(b) it follows that in the contact zone with the surface of the splat core, marked in dark gray in the figure on the right, pores are not formed due to significant discharge pressure when a drop of melt collides with the surface marked in black. Note that in Fig. 3(b) the right part of the splat periphery is not displayed, in contrast to the left part of the splat periphery shown in this figure, marked in a light gray shade (in the figure on the left). In addition, towards the periphery of the splat (in the figure on the

(a)   (b)

Fig. 3. Laying of the cylindrical splats (of gray shade) on a surface with a non-smoothed (variable) relief (black color): (a) – stepped relief; (b) – wave-shaped relief [6]. (Color figure online)

left), the pores increase. It should be noted that the splat core radius R_{c0} is determined by the formula $R_{c0} \cong 1.1R_p$, where R_p is the radius of the particle in the form of a spherical drop of melt.

The absence of pores between the splat core and the substrate is ensured by the first algorithm of the splat laying procedure – *the algorithm "for recognizing and filling depressions on the surface in the contact zone with the splat core"* [6]. The second algorithm in the splat laying procedure is *the algorithm "for recognizing and filling recesses behind obstacles on the surface in the contact zone with the peripheral annular part of the splat"* [6]. Next, a *wave algorithm* [6] is used *"for scanning the sprayed surface with a non-smoothed (variable) relief"* to prepare an array of "support vertices" with subsequent approximation of the lower surface of the splat using B-spline surfaces [7].

The splat laying simulation procedure uses discretization with certain steps Δx and Δy (usually $\Delta x = \Delta y$) in the XY plane. Discretization is implemented in the form of a two-dimensional grid of nodes corresponding to the elements of a two-dimensional array of "support vertices" [6]. The elements of these array represent the z-coordinates of the "support vertices" of the sprayed surface in "floating point arithmetic" (no discretization on the Z axis). After the above-mentioned filling of the recesses, the "residual" volume of the liquid melt of the spreading drop allows us to finally adjust the thickness h_s of the splat.

Based on the array of "support vertices," the spline approximation determines the z-coordinates of the splat lower surface, corresponding to the relief of the sprayed surface. Using the adjusted splat thickness h_s for the selected splat morphology (Fig. 2), the z-coordinates of the splat upper surface are calculated (the reliefs of both surfaces of the splat are similar to each other, Fig. 3). Both arrays of the z-coordinates of the lower and upper surfaces of the splat, corresponding to the x- and y-coordinates of the nodal points of the partition grid are stored on the external disk into the corresponding files. Using this file, in the "Functional characteristics of coatings" subsystem, coating characteristics (such as porosity $P\%$, roughness R_a, adhesive strength σ_{adg} of the coating to the base, etc.) are calculated.

3   Implementation of Program System for Parallel Running of SIMD-Tasks for Coating Simulation

3.1   Functionality of the Program System

As already mentioned in the introduction of the article, the creation of a program system with the ability to parallel execute SIMD-tasks is aimed at accelerating a series of simulations in which a certain class of coatings is modeled and their functional characteristics are calculated (porosity, strength of adhesion to the substrate, etc.).

The program system (PS) in terms of its functionality meets the following requirements:

- support for various formats for presenting and storing data;
- the ability to embed your own data presentation and storage formats;
- availability of a number of frequently used processing functions;
- the ability to create your own processing functions;
- the ability to develop, embed and execute various processing scenarios;
- convenient user interface with support for script debugging;
- display of processing results in a convenient form;
- system resistance to failures in processing functions;
- possibility of parallelization of calculations.

The PS implements the basic principle of optimal construction of software systems and complexes, which consists in creating several functional components (separate independent programs) interconnected by the ability to work with a common data storage (with files, databases, etc.). The PS structure is shown in Fig. 4.

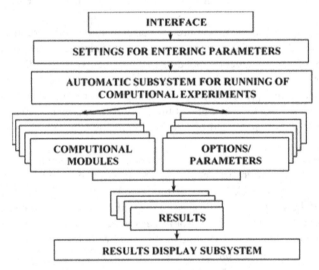

Fig. 4. Structure and composition of the PS.

The PS provides the following minimum set of components.

1. *Interface.* The main purpose of the interface is to enter initial data and parameters for executing a separate CE. The result of its work is a block of data of a certain structure, stored in a data warehouse, usually in the form of a specially designed file or record in a certain "database" (DB).
2. *Computing module (CM).* This program must access the data storage and find in it the corresponding "next" CE-block of input data corresponding to a certain coating simulation mode. Then the CM must perform basic calculations for simulating the coating with the specified parameters of the spraying mode and save the results in the

storage in the form of another data block. In the introduction to the article, it was noted that all CMs are identical software copies (Fig. 4), implementing the computational basis of the program complex for coating simulation (Fig. 1). The need to reduce the time spent on carrying out (both a separate CE and the entire cycle of CEs) forces the creation of a CM architecture in the form of identical copies of CMs independently executed from each other (executed in parallel) on different computing nodes. Many copies of CMs distributed on different computers (or on nodes of a computing cluster) should be considered as a whole single program that allows you to simultaneously (in parallel) carry out a certain number of CEs (or the entire CE-cycle) depending on the computing power of the used multiprocessor computing system (MCS). With this organization of the PS, "parallelism of applications (tasks)" is performed, which determines the SIMD-architecture of the PS.

3. *Module for processing calculation results.* This program uses the results obtained from the CM and processes them in order to obtain "visual" data for a researcher in the field of creating gas-thermal coatings. It would seem that such programs simply need an interface, but this is not always the case. In most cases, it is enough for the user to provide ready-made report files (electronic reports), which can be used in further work. Examples of such files can be files with graphs, or tables, images, and sometimes full "high availability" reports with which the researcher (user) works. These "high availability" electronic reports may not be "tied" to the interface and modules listed above, and must be opened with standard software (for example, MS Office or other programs) available on any PC to ensure smooth data exchange.

4. *Scripts for automatically performing computational experiments.* This block of programs deals with dispatching processes, sequentially and repeatedly run copies of CMs with different sets of input data of the same structure and sizes (run SIMD-tasks), waiting for the completion of SIMD-tasks. In addition, this block controls the results processing module, which can be run either immediately after the completion of each CM-copy, or after the completion of all CM-copies. Then it provides the generation of a general electronic report, simplifying further access to the data.

The block of scripts, in accordance with specified criteria, has the ability to partially edit input data received from the interface program, which allows you to automate data entry by the operator, as well as eliminate his errors. Scripts for automatically performing computational experiments are used to control the resources of computing nodes and running such a number of CM-copies that optimally uses "critical" resources. In the case when the number of computing nodes is less than the total number of CEs, each of the parallel running copies of the CM can execute several sequentially executed CEs.

The concept of experiment automation was a natural consequence of the development of coating simulation software [4–6]. This software allows you to calculate the functional and performance characteristics of powder coatings obtained by various gas-thermal technologies for spraying powder materials onto technical products (including APS, D-gun spraying and other methods). For a detailed study of the influence of various KPPs of sprayed particles and base (substrate) material on the characteristics of the coating, it was necessary to perform mass series of CEs.

Initially (before the development of the PS with an architecture for parallel running of similar SIMD-tasks), each CE was prepared manually: a set of parameters was specified for a specific CE through a graphical interface, then a calculation was carried out (lasting from several hours to several days). After completing the calculations of one CE, the next CE was again "manually" prepared and run for execution. Therefore, it was decided to move the sets of parameters into a separate configuration file and run the execution of all CEs in a special computing mode on a multiprocessor platform using the PS for parallel execution of SIMD-tasks.

The results of all CEs were saved to the PC hard drive. Further analysis of the results for each CE and their compilation into a general report "manually" sometimes led to errors. Therefore, it was decided to create a universal tabular report for each CE, including the KPPs of particles and substrate, which made it possible to further maintain a general log of all CEs, grouping them into series and recording time. Intermediate results reaching a volume of up to 20 GB for one CE are deleted after the completion of the report generation for each CE.

Detailed reports on the formation of coatings, as well as graphical versions of thin sections (vertical sections) of the coating, are of interest, but only if unexpected results are obtained as a result of the experiments. Despite the large volume of this data (about two gigabytes), it was decided to save the data on disk to restore the "evolutionary picture" of the experiments, including after a long time. To save disk space, software is used to reversibly compress information. Today, the volume of compressed results has already exceeded several terabytes, with an average compression ratio of ~80%, which suggests saving space on hard drives of at least ten terabytes.

Repeated modernization of the program complex for coating simulation gave rise to the problem of version control and changes, which they initially tried to solve using the traditional method, using the appropriate software (Hg Mercurial version control system). This made it possible to answer the question of exactly what changes were made, but did not allow us to compare the parameters and initial data of the CEs with the versions of the software package. Therefore, a rule was introduced: for each series of CEs, a separate current version of the source code of the software package is saved in order to subsequently track the changes made. Thus, the following algorithm was implemented as a set of scripts.

1. Initially, it is necessary to run the program and manually enter the parameters of the so-called "basic" CE spraying mode (technologically predictable spraying scenario).
2. Then the series name is assigned to the CEs series and the start date for the CEs series is indicated. Based on this, a separate directory is formed to store the results. A copy of the CM code is copied into the created directory, placed in the Src subdirectory and containing the basic parameter script.
3. In addition, many folders are created, the name of each of which contains a modified value of one of the KPPs, different from the value of this parameter from the "basic" set of KPPs (for example, Dp50 μm - a changed value of 50 μm for the particle diameter D_p instead of the value of 30 μm from the base set). A copy of the CM is copied to each such folder, as well as a configuration file with the corrected value of a specific parameter (the remaining parameters repeat the corresponding parameters of the previous set of KPPs). The created set of folders is copied to the computing nodes (computers), then, using scripts, programs are run from the corresponding folders for each CE (for each computing node).

4. At the end of the calculations, temporary files are deleted, the report file of the CEs cycle is copied to a shared folder, adding the results to the common "report database".

3.2 Computational Tests for Determination of Optimal Loading Mode of Computing Nodes

To identify optimal loading modes for computing nodes, a series of computational tests were carried out. The same computational tests were performed on three computing systems (on three samples of computing nodes) and the execution time of test calculations was studied. In different tests, different numbers of CM-copies were run, but each time the same number of copies was run for each of the three systems (see Table 1).

Table 1 shows the following time parameters: Time_1 is the time of the main calculations for the specified number of CM-copies for each of the three systems; Time_2 is the time for processing and analyzing the results obtained after running of the CM-copies (also for each of the three systems). The following computer system configurations (the samples of computing nodes) were used for testing:

- **System 1:** Intel Core i7-4710MQ (2.5 GHz): 4 cores, 8 computing threads, 8 GB RAM, Windows 7 x64, SSD: Kingston SV300S37A/240G (240 GB, communication speed of 450 MB/s);
- **System 2:** Intel Core i5-4690 (3.5 GHz): 4 cores, 4 computing threads, 12 GB RAM, Windows 7 x64, HDD: ST3500630AS (500 GB, spindle speed of 7200 rpm);
- **System 3:** Intel Core i5-4690 (3.5 GHz): 4 cores, 4 computing threads, 12 GB RAM, Windows 7 x64, SSD: Corsair Force GS (256 GB, communication speed of 500 MB/s).

As can be seen from Table 1, the optimal mode for each of the three systems is the use of 4 parallel computing processes (based on the time spent on executing one copy).

Table 1. Running time for multiple CM-copies on three computing systems.

Number of running CM-copies	Running time (in hours): Time_1 + Time_2 = Total time		
	System 1	System 2	System 3
1	3.3 + 1.1 = 4.4	4.0 + 1.5 = 5.5	3.6 + 1.2 = 4.8
2	5.6 + 1.9 = 7.5	6.5 + 2.7 = 9.2	6.0 + 2.1 = 8.1
3	6.1 + 2.5 = 8.6	7.3 + 3.4 = 10.7	6,8 + 2.7 = 9.5
4	7.0 + 1.6 = 8.6	7.9 + 2.1 = 10.0	7.6 + 1.7 = 9.3

The performance of each of the 3 systems was limited by hard drives. Therefore, CMs were run on high-speed hard drives in further running CEs. When distributing the load on computing nodes with different performance, the peculiarities of the influence of the parameter values of the initial sprayed particles on the coating calculation time were taken into account. The implemented architecture of the PS makes it possible to carry out parallel running CEs over a long period of time (in round-the-clock operation for several weeks). Upon completion of the work, a general electronic report is generated, which is duplicated in the archive.

4   Using a Program System with an Architecture for Parallel Running of Similar SIMD-Tasks for Coating Simulation

Analyzing the simulation results of coatings made of titanium nickelide (TiNi), taking into account various sets of KPPs characterizing different modes of coating spraying, it can be stated that the following functional characteristics (properties) of coatings change significantly: R_a – coating roughness; $P\%$ – coating porosity; σ_{adg} – strength of adhesion of the coating to the substrate.

The main results of the CEs at the stage of modeling coatings (with thickness $H = 100\ \mu$m) are summarized in Tables 2, 3, 4, 5, 6, 7, 8 and 9, in which, similarly to work [8], in addition to the relative adhesive strength of coatings $\overline{\sigma}_{adg}$, coating roughness R_a and coating porosity $P\%$, the powder particle KPPs and calculated splat parameters are given. Similarly to work [8], in Tables 2, 3, 4, 5, 6, 7, 8 and 9 the following designations of quantities are used: T_p – particle temperature; U_p – particle velocity; D_p – particle diameter; T_b – base (substrate) temperature; D_s – splat diameter; h_s – splat height (thickness at its center).

Similarly to work [8], the value $\overline{D}_s$ is normalized value of the splat diameter D_s, expressed in relative units (r.u.) and which is otherwise called the "spreading factor" of the melt particle ($\overline{D}_s = D_s/D_p$). The value $\overline{h}_s$ is the normalized value of the splat thickness h_s, i.e. $\overline{h}_s = h_s/D_p$. The value $\overline{\sigma}_{adg}$ determines the relative strength of adhesion of the coating to the base (more precisely, of the first monolayer of the coating to the substrate). In other words, it determines the ratio of the absolute adhesive strength of the coating to the tensile strength (i.e., it determines the proportion of the maximum possible adhesive strength). Similarly to work [8], the temperature T_c determines the contact temperature at the point of collision of the melt droplet with the surface (with the substrate). The integer value of the "Scenario" parameter indicates the number of the scenario of the melt droplet spreading on the substrate (see below) [8, 9]:

1. spreading and simultaneous solidification of the droplet on the solid base;
2. spreading and simultaneous solidification of the droplet, and local submelting of the base at the contact spot with the droplet;
3. spreading of the droplet over the solid base surface, and subsequent cooling and solidification of the spread layer;
4. spreading of the droplet accompanied with simultaneous local submelting of the base, followed by subsequent cooling and solidification of both.

Let's analyze Tables 2, 3, 4, 5, 6, 7, 8 and 9, in which, based on the results of the CEs, the calculated splat parameters and functional characteristics of TiNi coatings on steel (Steel45) and titanium (Ti) substrates are entered.

The value $\overline{D}_s$ (spreading factor) for splats from metal [10] must satisfy the condition $\overline{D}_s < 4.5$ (starting from a value of 5, breaks appear on the circular periphery of the splat). As can be seen from Tables 2, 3, 4, 5, 6, 7, 8 and 9, this condition is satisfied

Table 2. Calculated splat parameters and functional characteristics of TiNi coatings on a Steel45 substrate when changing the temperature parameter T_p of the particles and maintaining the conditions: $U_p = 200$ m/s, $D_p = 30$ μm, $T_b = 400$ K.

T_p, K	D_s, μm	$\overline{D}_s$, r.u	h_s, μm	$\overline{h}_s$, r.u	R_a, μm	$P\%$	$\overline{\sigma}_{adg}$, r.u	T_c, K	Scenario
1600	119	3.950	1.28	0.043	1.98	3.998	0.0257	1222.4	1
1650	121	4.038	1.23	0.041	1.92	3.936	0.0431	1256.7	1
1700	150	5.002	0.81	0.027	2.09	3.942	0.0967	1291.0	1
1750	201	6.701	0.45	0.015	2.05	4.239	0.1521	1325.2	3
1800	201	6.701	0.45	0.015	1.99	4.009	0.1997	1359.5	3
1850	201	6.701	0.45	0.015	2.04	3.981	0.2475	1393.8	3

Table 3. Calculated splat parameters and functional characteristics of TiNi coatings on a Steel45 substrate when changing the velocity parameter U_p of the particles and maintaining the conditions: $T_p = 1600$ K, $D_p = 30$ μm, $T_b = 400$ K.

U_p, m/s	D_s, μm	$\overline{D}_s$, r.u	h_s, μm	$\overline{h}_s$, r.u	R_a, μm	$P\%$	$\overline{\sigma}_{adg}$, r.u	T_c, K	Scenario
100	100	3.337	1.80	0.060	1.63	11.513	0.0238	1122.0	1
150	110	3.682	1.48	0.049	1.56	10.184	0.0242	1122.0	1
200	119	3.950	1.28	0.043	1.98	3.998	0.0257	1222.4	1
250	210	7.007	0.41	0.014	2.34	1.685	0.0261	1222.4	3
300	218	7.267	0.38	0.013	2.10	0.690	0.0280	1222.4	3
350	225	7.494	0.36	0.012	0.48	0.766	0.0277	1222.4	3
400	231	7.697	0.34	0.011	0.46	0.896	0.0245	1222.4	3

by scenario 1, and scenario 3 is not satisfactory. When spraying TiNi particles onto a titanium substrate, only scenario 3 is practically realized, with the exception of the technological spraying mode (scenario 1, $\overline{D}_s < 4.5$): $T_p = 1600$ K, $U_p = 100$ m/s, $D_p = 20$ μm, $T_b = 400$ K. Therefore, from a technological point of view, spraying coatings of titanium nickelide particles onto a steel substrate deserves more attention. The technological mode of spraying coatings of TiNi particles onto a Steel45 substrate corresponds to wider ranges of KPPs, namely: $T_p = 1600$–1700 K, $U_p = 100$–200 m/s, $D_p = 20$–30 μm, $T_b = 300$–500 K.

It is known [11] that with increasing particle temperature T_p, and consequently with increasing contact temperature T_c, the relative adhesive strength $\overline{\sigma}_{adg}$ initially increases rapidly, and then the increase slows down (and may approach 1). The above mentioned growth trend of $\overline{\sigma}_{adg}$ is conformed in Table 2. The contact temperature T_c also increases with increasing substrate temperature T_b (see Table 5), therefore the relative adhesive strength $\overline{\sigma}_{adg}$ also increases (up to approximately 0.045). However, as the

Table 4. Calculated splat parameters and functional characteristics of TiNi coatings on a Steel45 substrate when changing the diameter parameter D_p of the particles and maintaining the conditions: $T_p = 1600$ K, $U_p = 200$ m/s, $T_b = 400$ K.

D_p, μm	D_s, μm	$\overline{D}_s$, r.u	h_s, μm	$\overline{h}_s$, r.u	R_a, μm	$P\%$	$\overline{\sigma}_{adg}$, r.u	T_c, K	Scenario
20	72	3.578	1.04	0.052	1.21	3.432	0.0179	1122.0	1
30	119	3.950	1.28	0.043	1.98	3.998	0.0257	1222.4	3
40	284	7.098	0.53	0.013	2.13	4.529	0.0446	1222.4	3
50	371	7.422	0.61	0.012	2.62	5.070	0.0627	1222.4	3
60	462	7.697	0.68	0.011	3.89	5.851	0.0822	1222.4	3
70	556	7.938	0.74	0.011	7.12	8.283	0.0994	1222.4	3
80	652	8.153	0.80	0.010	8.46	10.422	0.1191	1222.4	3
90	751	8.347	0.86	0.010	12.24	13.526	0.1402	1222.4	3
100	853	8.525	0.92	0.009	15.07	16.667	0.1577	1222.4	3

Table 5. Calculated splat parameters and functional characteristics of TiNi coatings on a Steel45 substrate when changing the base (substrate) temperature parameter T_b and maintaining the conditions: $T_p = 1600$ K, $U_p = 200$ m/s, $D_p = 30$ μm.

T_b, K	D_s, μm	$\overline{D}_s$, r.u	h_s, μm	$\overline{h}_s$, r.u	R_a, μm	$P\%$	$\overline{\sigma}_{adg}$, r.u	T_c, K	Scenario
300	116	3.852	1.35	0.045	1.97	3.912	0.0148	1191.0	1
400	119	3.950	1.28	0.043	1.98	3.998	0.0257	1222.4	1
500	122	4.061	1.21	0.040	2.05	4.126	0.0450	1253.9	1

particle temperature T_p increases, the relative adhesive strength $\overline{\sigma}_{adg}$ reaches a higher value of 0.2475.

As can be seen from Table 4, with increasing diameter D_p of particles, the value $\overline{\sigma}_{adg}$ also increases (to a value of 0.1577), giving way to growth of $\overline{\sigma}_{adg}$ with increasing of T_p.

According to the conclusions given in [11], the relative adhesive strength $\overline{\sigma}_{adg}$ in theoretical terms does not change when the velocity of particles changes (see Table 3).

From Table 2 it follows that for an acceptable technological mode with the highest value $\overline{\sigma}_{adg} = 0.0967$ can be corresponded the mode with the parameters: $T_p = 1700$ K, $U_p = 200$ m/s, $D_p = 30$ μm, $T_b = 400$ K. Table 5 indicates an increase of $\overline{\sigma}_{adg}$ of approximately 1.751 times as the substrate temperature increases from 400 to 500 K (while maintaining scenario 1 and $\overline{D}_s < 4.5$). Therefore, we can consider the optimal technological spraying mode to be the one with the following parameters: $T_p = 1700$ K, $U_p = 200$ m/s, $D_p = 30$ μm, $T_b = 500$ K, for which the estimate $\overline{\sigma}_{adg}$ will increase by 1.751 times, i.e. $\overline{\sigma}_{adg} \cong 0.17$.

Table 6. Calculated splat parameters and functional characteristics of TiNi coatings on a titanium substrate when changing the temperature parameter T_p of the particles and maintaining the conditions: $U_p = 200$ m/s, $D_p = 30$ μm, $T_b = 400$ K.

T_p, K	D_s, μm	$\overline{D}_s$, r.u	h_s, μm	$\overline{h}_s$, r.u	R_a, μm	$P\%$	$\overline{\sigma}_{adg}$, r.u	T_c, K	Scenario
1600	201	6.701	0.45	0.015	1.72	3.880	0.0006	1222.4	3
1650	201	6.701	0.45	0.015	1.90	4.030	0.0011	1256.7	3
1700	201	6.701	0.45	0.015	1.96	4.148	0.0020	1291.0	3
1750	201	6.701	0.45	0.015	2.05	3.984	0.0029	1325.2	3
1800	201	6.701	0.45	0.015	1.89	4.022	0.0048	1359.5	3
1850	201	6.701	0.45	0.015	2.00	4.244	0.0078	1393.8	3

Table 7. Calculated splat parameters and functional characteristics of TiNi coatings on a titanium substrate when changing the velocity parameter U_p of the particles and maintaining the conditions: $T_p = 1600$ K, $D_p = 30$ μm, $T_b = 400$ K.

U_p, m/s	D_s, μm	$\overline{D}_s$, r.u	h_s, μm	$\overline{h}_s$, r.u	R_a, μm	$P\%$	$\overline{\sigma}_{adg}$, r.u	T_c, K	Scenario
100	111	3.705	1.46	0.049	1.76	11.606	0.0002	1122.0	1
150	190	6.326	0.50	0.017	1.83	10.423	0.0009	1122.0	3
200	201	6.701	0.45	0.015	1.72	3.880	0.0006	1222.4	3
250	210	7.007	0.41	0.014	2.32	1.652	0.0005	1222.4	3
300	218	7.267	0.38	0.013	2.19	0.728	0.0004	1222.4	3
350	225	7.494	0.36	0.012	0.42	0.774	0.0004	1222.4	3
400	231	7.697	0.34	0.011	0.45	0.895	0.0003	1222.4	3

In [12–14] for titanium nickelide (B2-phase of TiNi) data are given on the tensile strength (denoted as σ_{ts}). In [12] states that the presence of phase precipitates Ti_2Ni and $TiNi_3$ in the B2-phase matrix of TiNi, even with small concentrations, leads to exceeding the tensile strength value of 950–1000 MPa, which characterizes the minimum value σ_{ts} for the B2-phase of TiNi. In [13] a high tensile strength value of 1885 MPa was achieved for wire made of nanostructured titanium nickelide, taking into account multiple compression by drawing and controlled multi-stage heat treatment. In [14] on page 21 indicates that with a certain chemical composition of titanium nickelide, as well as during thermomechanical processing, the tensile strength can reach values of up to 2000 MPa. A value of 1885 MPa can be taken as the base value for the B2-phase of TiNi.

Table 8. Calculated splat parameters and functional characteristics of TiNi coatings on a titanium substrate when changing the diameter parameter D_p of the particles and maintaining the conditions: $T_p = 1600$ K, $U_p = 200$ m/s, $T_b = 400$ K.

D_p, μm	D_s, μm	$\overline{D}_s$, r.u	h_s, μm	$\overline{h}_s$, r.u	R_a, μm	$P\%$	$\overline{\sigma}_{adg}$, r.u	T_c, K	Scenario
20	79	3.974	0.84	0.042	1.31	3.414	0.00006	1122.0	1
30	201	6.701	0.45	0.015	1.72	3.880	0.00057	1222.4	3
40	284	7.098	0.53	0.013	2.20	4.665	0.00086	1222.4	3
50	371	7.422	0.61	0.012	3.11	5.019	0.00124	1222.4	3
60	462	7.697	0.68	0.011	4.09	6.047	0.0010	1222.4	3
70	556	7.938	0.74	0.011	7.81	8.859	0.0011	1222.4	3
80	652	8.153	0.80	0.010	8.30	10.399	0.0017	1222.4	3
90	751	8.347	0.86	0.010	11.83	14.143	0.0019	1222.4	3
100	853	8.525	0.92	0.009	14.51	17.110	0.0018	1222.4	3

Table 9. Calculated splat parameters and functional characteristics of TiNi coatings on a titanium substrate when changing the base (substrate) temperature parameter T_b and maintaining the conditions: $T_p = 1600$ K, $U_p = 200$ m/s, $D_p = 30$ μm.

T_b, K	D_s, μm	$\overline{D}_s$, r.u	h_s, μm	$\overline{h}_s$, r.u	R_a, μm	$P\%$	$\overline{\sigma}_{adg}$, r.u	T_c, K	Scenario
300	201	6.701	0.45	0.015	2.01	3.964	0.00032	1191.0	3
400	201	6.701	0.45	0.015	1.72	3.880	0.00057	1222.4	3
500	201	6.701	0.45	0.015	1.93	4.006	0.00099	1253.9	3

The "absolute" adhesive strength σ_{adg} is determined by multiplying the relative adhesive strength $\overline{\sigma}_{adg}$ by the "tensile" strength value σ_{ts}, i.e. [11]

$$\sigma_{adg} = \overline{\sigma}_{adg} \cdot \sigma_{ts}.$$

Using the more likely value range of 1885–2000 MPa, corresponding to the σ_{ts}-values (values decreasing towards 1000 MPa are unlikely) and the estimate $\overline{\sigma}_{adg} = 0.17$ (obtained above), the absolute adhesive strength values for TiNi coatings range from 320 to 340 MPa (a range 170–320 MPa is unlikely). We can say that the obtained estimates of the expected range of adhesive strength for TiNi coatings on a steel substrate are quite high.

For acceptable spraying modes, in particular for the optimal technological mode ($T_p = 1700$ K, $U_p = 200$ m/s, $D_p = 30$ μm, $T_b = 500$ K), the roughness R_a is quite small (about 2 μm) and is less than for many other coatings [15]. Its share in relation to the thickness of the coatings ($H = 100$ μm) is about 2%.

Analyzing the porosity of TiNi coatings on both steel and titanium substrates, we can say that the porosity $P\%$ of the coatings practically varies little when the particle temperature T_p changes within 250 K (see Tables 2 and 6). The porosity of TiNi coatings also varies little for both types of substrate and with changes in substrate temperature T_b (see Tables 5 and 9). Depending on the increase in particle diameter D_p (Tables 4 and 8), porosity increases significantly (close to a linear trend), which is typical for many metal coatings. But within the range of changes in particle diameter up to 70 μm (see Tables 4 and 8), porosity remains within the range of 4 to 8%, traditional for metal coatings in the case of APS. The particle velocity for the APS spraying mode is in the range of 200–300 m/s. An expected fact is that porosity tends to decrease with increasing particle velocity (see Tables 3 and 7). Starting from values of 300 m/s and higher (for example, for D-gun spraying and HVOF technologies), the porosity decreases sharply and becomes less than 1%, thereby realizing very dense coatings that are in demand in the materials industry [16–20].

5 Conclusions

Summarizing the results of computational experiments, we can say that the adhesive strength of TiNi coatings is quite high and practically does not depend on particle velocity. In plasma spraying mode (APS), the porosity of coatings does not exceed traditional values (up to 7–8%), and for D-gun spraying and HVOF – less than 1%. The roughness of TiNi coatings is quite small and does not exceed 2 μm. Such very dense coatings with low roughness are in demand in the materials industry [16–20]. Formulating conclusions regarding the created modeling software, we can say the following.

The implemented program system, despite its simplicity, provides a number of significant advantages:

- there is no need to parallelize each copy of the CM;
- systematization of the obtained CE-results is carried out, the ability to quickly create a general electronic report with illustrative research material;
- saving time by automating the preparation of a general electronic report (instead of the work of the researcher to compile the report "manually");
- the CEs cycle is carried out in a "conveyor" way and there is no loss of time between the execution of individual CEs;
- errors arising due to the so-called "human factor" are practically eliminated;
- a compact archive is created that makes it possible to obtain comprehensive data about each CE after a long time, which fully justifies the time spent on developing the corresponding software modules.

This technique has been tested and is recommended for embedding into computer programs that perform a series of similar calculations.

Acknowledgments. This study was supported by the Basic Research Program of the State Academies of Sciences for 2021–2023, project no. AAAA-A17-117030610124-0. This study was performed on the equipment of the Collective Use Center Mechanics (Khristianovich Institute of Theoretical and Applied Mechanics of the Siberian Branch of the Russian Academy of Sciences).

References

1. Kudinov, V.V., Bobrov, G.V.: Spray Coating Deposition. Theory, Technology and Equipment, 432 p. Metallurgiya, Moscow (1992). (in Russian)
2. Kalita, V.I., Komlev, D.I.: Plasma Coatings with Nanocrystalline and Amorphous Structure: Monograph, 388 p. Lider M, Moscow (2008). (in Russian)
3. Alkhimov, A.P., Klinkov, S.V., Kosarev, V.F., Fomin, V.M.: Cold gas-dynamic spraying. Theory and practice. Ed. V.M. Fomin, 536 p. Fizmatlit, Moscow (2010). (in Russian)
4. Solonenko, O.P., Blednov, V.A., Iordan, V.I.: Computer design of thermal sprayed metal powder coatings. Thermophys. Aeromech. **18**(2), 255–272 (2011). https://doi.org/10.1134/s0869864311020065
5. Solonenko, O.P., Jordan, V.I., Blednov, V.A.: Stochastic computer simulation of cermet coatings formation. Adv. Mater. Sci. Eng. **2015**, 396427 (2015). https://doi.org/10.1155/2015/396427
6. Jordan, V.I., Blednov, V.A.: Computer-aided design and prediction of the functional properties of double-layered thermal barrier coatings NiAl-YSZ. IOP Conf. Ser.: J. Phys.: Conf. Ser. **1134**, 012024 (2018). https://doi.org/10.1088/1742-6596/1134/1/012024, https://iopscience.iop.org/article/10.1088/1742-6596/1134/1/012024/pdf
7. Jordan, V.I., Blednov, V.A., Solonenko, O.P.: B-spline surfaces in the algorithm for adapting the shape of splats to the changing topology of the sprayed coating surface when simulating the formation of its structure. In: Proceedings of the IX All-Russian scientific-practical conference with international participation "Information technologies and mathematical modeling (ITMM-2010)", Tomsk, Russia, 19–20 November 2010, vol. 2, pp. 155–160. Tomsk State University (2010). (in Russian)
8. Jordan, V.I., Blednov, V.A., Smirnov, A.V., Chesnokov, A.E.: Study of influence of sprayed particles parameters on adhesion strength of copper coatings using computer simulation methods. In: AIP Conference Proceedings of the Conference "Actual problems of continuum mechanics: experiment, theory, and applications", Novosibirsk, Russia, 20–24 September 2021, vol. 2504, no. 1, p. 030019. AIP Publishing (2023). https://doi.org/10.1063/5.0132440. https://doi.org/10.1063/5.0132440
9. Solonenko, O.P.: State-of-the art of thermophysical fundamentals of plasma spraying. In: Solonenko, O.P., Zhukov, M.F. (eds.) Thermal Plasma and New Materials Technology, vol. 2, pp.7–96. Cambridge Interscience Publishing, Cambridge (1995)
10. Solonenko, O.P., Smirnov, A.V.: Equilibrium solidification of melted microdroplets under their collision with substrate: model experiment and criterial generalization of splats morphology. In: Thermal Plasma Torches and Technologies, vol. 2, pp. 99–113. Cambridge International Science Publishing, Cambridge (2001)
11. Kudinov, V.V.: Plasma Coatings, 184 p. Nauka, Moscow (1977). (in Russian)
12. Soldatova, M.I., Khodorenko, V.N., Gunter, V.E.: Physico-mechanical and strength properties of alloys based on titanium nickelide (TN-10, TN-20, TN-1V). News of Tomsk Polytech. Univ. Math. Mech. **322**(2), 135–139 (2013). (in Russian)
13. Baikin, A.S.: Development of a composite biomedical material "Nanostructural titanium nickelide - a biodegradable polymer". Author's abstract of Ph.D. Dissertation. Institute of Metallurgy and Materials Science named after. A. A. Baykov, Russian Academy Sciences (IMET RAS), Moscow (2019). (in Russian)
14. Vinogradov, R.E.: Thermo-mechanical behavior of functional metal-polymer composite materials reinforced with titanium nickelide. Ph.D. Dissertation. National research university "Moscow Aviation Institute", Moscow (2022). (in Russian)
15. Lebedev, D.I., Vinokurov, G.G., Struchkov, N.F.: Open porosity and roughness of the surface of wearproof powder coatings. Acad. J. "Izv. Samara Sci. Cent. Russ. Acad. Sci." **20**(6), 68–72 (2018). (in Russian)

16. Ulianitsky, V.Yu., Shtertser, A.A., Batraev, I.S.: Electrical insulating properties of aluminum oxide detonation coatings. Metal Process. (Technol. Equip. Tools) **20**(4), 83–95 (2018). https://doi.org/10.17212/1994-6309-2018-20.4-83-95, https://journals.nstu.ru/files/articles/flash/18007/83/#zoom=z. (in Russian)
17. Belotserkovskiy, M.A., Levantsevich, M.A., Konovalova, E.F.: Surface modification of thermal spraying stainless steel coatings by cladding deformation. Mech. Mach. Mech. Mater. **1**(34), 64–67 (2016). (in Russian)
18. Yurkinsky, V.P., Firsova, E.G., Okovity, V.V.: The effect of carbon steel (types steel c1020) and stainless steel (types steel 321) oxidation technique on the porosity of oxide coatings. Sci. Tech. Statements St. Petersburg State Polytech. Univ. **2**(171), 133–137 (2013). (in Russian)
19. Tretyakov, A.F.: Design technique of manufacturing process of porous material products with desired properties. Eng. J.: Sci. Innov. **2**, 1–15 (2017). https://doi.org/10.18698/2308-6033-2017-02-1588. (in Russian)
20. Fen', E.К.: Deposition of coverings based on composite powder materials by detonation method. Prod. Tech. J. "Welder" **1**(59), 17–19 (2016). (in Russian)

On Hierarchical Convergence of the Heterogeneous Multiscale Finite Element Method Using Polyhedral Supports

Anastasia Yu. Kutishcheva[✉] [iD], Sergey I. Markov[iD], and Ella P. Shurina[iD]

Trofimuk Institute of Petroleum Geology and Geophysics, SB RAS, Koptug ave. 3, 630090 Novosibirsk, Russia
KutischevaAY@ipgg.sbras.ru

Abstract. Numerical simulation of physical processes in such complex formations as permafrost is impossible without modern multiscale methods. For example, the heterogeneous multiscale finite element method (FE-HMM) can be used to simulate elastic deformation of solids. However, it is necessary to have some control tools of computational errors as well as a priori information on limitations of the method in solving practical problems. In this article, we investigate the influence of different mesh hierarchy levels on the accuracy and speed of solving the elastic deformation problem using FE-HMM with polyhedral supports at the macrolevel and with tetrahedral supports at the microlevel. The obtained estimates allow us to adjust the planning strategies of computational experiments, which are largely related to the construction of a set of meshes with different accuracy. We apply the h-refinement technology at the edges of macro-polyhedra in the microlevel mesh. Increasing in accuracy of computational solution up to two orders of magnitude with maintaining the total size of discretizations is obtained. Also, the applicability of the method for numerical simulation of physical processes in media with elongated inhomogeneities intersecting several macroelements is shown.

Keywords: Elastic Deformation · Heterogeneous Multiscale Finite Element Method · Polyhedral Supports · Natural Parallelism

1 Introduction

Structural complexity of rocks imposes certain limitations on the computational schemes used. For example, direct methods (such as the Galerkin method [1]) lead to discretizations with a large number of parameters. It is critical situation even using modern computing machines. Therefore, methods based on decomposition of the initial domain into some subdomains are applied to simulate physical processes in heterogeneous multiscale media. However, in this case it is necessary to construct special boundary operators to ensure the continuity of the solution at interfaces. The most general and easy-to-use mathematical tools for constructing such operators are offered by multiscale finite element methods. For example, in the heterogeneous multiscale finite element method

© The Author(s), under exclusive license to Springer Nature Switzerland AG 2024
V. Jordan et al. (Eds.): HPCST 2023, CCIS 1986, pp. 92–104, 2024.
https://doi.org/10.1007/978-3-031-51057-1_7

(FE-HMM) [2–4], the solution space has a hierarchical structure and contains two or more levels. Each of levels corresponds to one physical or geometric scale of the problem under consideration. At each level of the hierarchy, special subproblems are formulated to construct local subspaces. The procedures for matching the hierarchy levels are chosen according to the specifics of the processes being simulated. This strategy ensures sufficient flexibility and adaptability of the method.

In this paper, we recall the upper mesh level as a macrolevel taking into account the effective properties of the medium. We recall the lower mesh level as a microlevel. The microlevel allows to take into account all inhomogeneities of the medium. To simulate the elastic deformation of a heterogeneous solid using the two-level FE-HMM, let us formulate the main stages:

1. Construction of the mesh hierarchy: partitioning of the entire modelling domain into subdomains (macroelements) without overlapping and without strictly considering the internal geometric structure. Then each of the macroelements is independently partitioned into microelements, taking into account the internal structure.
2. Construction of a functional hierarchy. To construct the macroscale shape functions and to ensure smoothness of the solution, we consider series of subproblems on each macroelements using microelement meshes.

Despite the sufficient representation of FE-HMM in the literature, including for modelling elastic deformation [5], some mathematical and corresponding technological aspects of this method should be developed for each specific physical problem. In addition, at the upper level of the hierarchy (hereinafter referred to as the macrolevel) finite element of a simple shape with an equal number of faces (tetrahedrons or parallelepipeds) are traditionally used to simplify the construction of functional subspaces of the solution. However, to simulate the elastic deformation of heterogeneous solids, it is more efficient to use polyhedral macroelements with a non-fixed number of faces. We have proposed and implemented technologies for automating construction of mesh discretizations at the macrolevel, and approaches to construct the macroscale shape functions ensuring the required accuracy of the numerical solution.

2 Problem Statement and Solution Method

Let $\Omega = [0, 1] \times [0, 1] \times [0, 1]$ be a homogeneous modelling domain. We consider a sandstone with Young's modulus 5 GPa and Poisson's ratio 0.27 GPa.

The lower base of the area is rigidly fixed. We assume that the upper base is displaced upward along the Z axis by 0.1 m. The side surface is free. The force of gravity is not taken into account in this case. Thus, the boundary elastic deformation problem is presented as:

$$-\nabla \cdot (\mathbf{D} : \nabla_s \mathbf{U}(\mathbf{x})) = 0, \tag{1}$$

$$\mathbf{U}|_{lower\_base} = (0, 0, 0)^T, \tag{2}$$

$$\mathbf{U}|_{upper\_base} = (0, 0, 0.1)^T, \tag{3}$$

$$\mathbf{n} \cdot \sigma|_{side\_surface} = (0, 0, 0)^{T}. \tag{4}$$

If the domain is inhomogeneous, the above formulation is valid, but the corresponding ideal contact conditions at the interfaces are added.

2.1 Discretization of Computational Domain

This work does not deal with the construction of adaptive tetrahedral meshes in complex three-dimensional domains. For this purpose, the open integrated numerical modelling platform SALOME is used. It provides a wide range of tools for working with geometric objects and with mesh structures.

Polyhedral meshes are constructed as duals to the primary tetrahedral mesh. The nodes of the polyhedral mesh are the barycenters of geometric elements of the primary mesh. This approach is based on the Voronoi diagram [6], which is dual to the Delaunay triangulation [7]. There are different realizations of this approach in the two-dimensional and three-dimensional domains [8–12]. The approach is usually related with the choice of the way of constructing boundary elements, since in the basic formulation, the Voronoi diagram is constructed in the infinite domain.

Let $T(\Omega) = \left\{ T^{3D}, T^{2D}, T^{1D}, T^{0D} \right\}$ be some primary tetrahedral mesh constructed in a single-connected domain Ω with the non-smooth external boundary $\partial\Omega$.

The $T(\Omega)$ consists of a subset of tetrahedra $T^{3D} = \left\{ t_i^{3D}, i = \overline{1, n_{T3D}} \right\}$, faces $T^{2D} = \left\{ t_i^{2D}, i = \overline{1, n_{T2D}} \right\}$, edges $T^{1D} = \left\{ t_i^{1D}, i = \overline{1, n_{T1D}} \right\}$, and nodes $T^{0D} = \left\{ t_i^{0D}, i = \overline{1, n_{T0D}} \right\}$. We recall $\Pi(\Omega) = \left\{ \Pi^{3D}, \Pi^{2D}, \Pi^{1D}, \Pi^{0D} \right\}$ as a polyhedral mesh. We have implemented the most general approach, which consists of the sequential processing of the following basic mappings:

$$V_{0D} : \left(T^{0D} \right) \mapsto \Pi^{3D}, \tag{5}$$

$$F_{1D} : \left(T^{1D} \right) \mapsto \Pi^{2D}, \tag{6}$$

$$E_{2D} : \left(T^{2D} \right) \mapsto \Pi^{1D}, \tag{7}$$

$$N_{3D} : T^{3D} \mapsto \Pi^{0D}, \tag{8}$$

where V_{0D} is mapping, which corresponds to a node of the primary mesh to a polyhedron; F_{1D} is mapping, which corresponds the edge of the primary mesh element to the polyhedral mesh edge (Fig. 1a); E_{3D} is mapping, which corresponds to the inner edge of the primary mesh element to the polyhedral mesh edge (Fig. 1b); N_{3D} is mapping, matching the barycenters of the primary mesh elements to the polyhedral mesh nodes.

In addition, external boundaries $\partial\Omega$ give rise to additional special cases. For example, we need have a mapping that maps a primary mesh node lying on an external boundary $\partial\Omega$ to a polyhedral mesh edge (Fig. 2):

$$F_{0D} : \left\{ t^{0D} | t^{0D} \in T^{0D} \; \exists t^{2D} \in T_{ext}^{2D} : \; t^{2D} \cap t^{0D} = t^{0D} \right\} \mapsto \Pi_{0D}^{2D}, \tag{9}$$

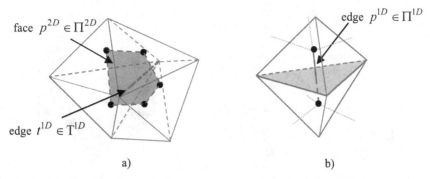

Fig. 1. Action of mapping operators from an edge of the primary mesh to a polyhedral mesh edge (a) and mapping from a primary mesh edge to a polyhedral mesh edge (b)

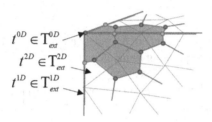

Fig. 2. Action of the mapping operator from a primary mesh node lying on the external boundary $\partial \Omega$ to a polyhedral mesh edge

The considered approach of polyhedral mesh construction has some disadvantages, such as the possibility of non-planar faces and non-convex polyhedra. However, for the computational schemes based on multiscale methods proposed in this paper, this is not a critical situation.

2.2 Heterogeneous Multiscale Finite Element Method

We consider a polyhedral consistent irregular mesh $\Pi^H(\Omega) = \left\{ K_i^{\tilde{p}}, \ i = \overline{1, N} \right\}$ of the entire computational domain. Where $K_i^{\tilde{p}}$ is a nonoverlapping polyhedron (Fig. 3a), and the $\tilde{p}$ is the number of the polyhedron vertices.

Let us introduce a finite-dimensional subspace on the set $\Pi^H(\Omega)$:

$$\left[L^2(\Omega)\right]^3 \supset \mathbf{V}^H = \left\{ \mathbf{u} | \mathbf{u} \in \left[L^2(\Omega)\right]^3 : \mathbf{u} \in \left[\Im^H\left(\Pi^H(\Omega)\right)\right]^3 \right\}, \qquad (10)$$

where $\Im^H\left(\Pi^H(\Omega)\right) = \left\{ \mathbf{\Psi}_p, p = \overline{1, P} \right\}$ is the space of non-polynomial finite shape functions associated with mesh nodes $p = \overline{1, P}$. Since the functions $\mathbf{\Psi}_p$ are finite and satisfy the "unit partitioning"-condition by the FE-HMM requirements [13], we construct consistent irregular tetrahedral meshes $T^h\left(K_i^{\tilde{p}}\right) = \left\{ \tau_j, \ j = \overline{1, n_i} \right\}, \forall i = \overline{1, N}$ (see Fig. 3b)

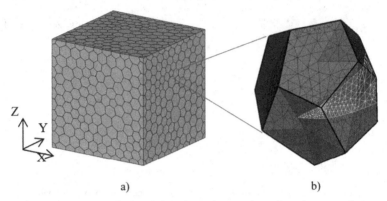

Fig. 3. Structure of the computational domain and examples of meshes: a – the macromesh $\Pi^H(\Omega)$ consists of 1290 polyhedra; b – the micromesh $T^h\left(K_i^{\tilde{p}}\right)$ in the polyhedron

on each of the polyhedral macroelement $K_i^{\tilde{p}}$ independently. To construct them, we define the next subspaces on corresponding finite element:

$$\left[L^2(\Omega)\right]^3 \supset \mathbf{V}^h = \left\{\boldsymbol{\Psi} \,|\, \boldsymbol{\Psi} \in \left[L^2(\Omega)\right]^3 : \boldsymbol{\Psi} \in \left[\Upsilon^h(\tau_j)\right]^3 \forall \tau_j \in T^h(\Omega)\right\}, \qquad (11)$$

where $\Upsilon^h(\tau_j)$ is a space of first-degree polynomials (in the general case polynomials have the degree m). Then let us formulate discrete variational formulations at the microlevel for $K_i^{\tilde{p}}$, $\forall p = \overline{1, \tilde{p}}$, $\forall i = \overline{1, N}$:

$$\begin{cases} \text{find } \boldsymbol{\Psi}_p^h \in \mathbf{V}^h + \widetilde{\boldsymbol{\Psi}}_p\left(\partial K_i^{\tilde{p}}\right) \text{ such that} \\ \int\limits_{\Omega} \nabla \boldsymbol{\Psi}_p^h : \mathbf{D} : \nabla \mathbf{v}^h d\Omega = 0, \ \forall \mathbf{v}^h(\mathbf{x}) \in \mathbf{V}^h, \end{cases} \qquad (12)$$

where $\widetilde{\boldsymbol{\Psi}}_p\left(\partial K_i^{\tilde{p}}\right)$ is the value of the function $\boldsymbol{\Psi}_p^h$ on the external boundary of the polyhedron. The function can be found from the solution of a similar problem with the lower dimensionality. Thus, we need to solve the quasi-dimensional problem on the boundary (Fig. 4b). Also, the values of the desired function along the edges are required (Fig. 4c). They can be obtained by using the simple interpolation from $(0, 0, 0)^T$ to $(1, 1, 1)^T$. If the edge is not crossed by an inclusion, in this case we construct one-dimensional subproblems along the edges with the boundary conditions $(0, 0, 0)^T$ or $(1, 1, 1)^T$ depending on the function index. Thus, the weight of the resulting function $\boldsymbol{\Psi}_p^h$ and degree-of-freedom corresponding to the vertex with the index p equals to the value $(1, 1, 1)^T$. All other weights associated with the macroelement vertices are equal to $(0, 0, 0)^T$.

Also, by definition FE-HMM the shape functions must fulfill the property (Fig. 5):

$$\sum_{p=1}^{\tilde{p}} \boldsymbol{\Psi}_p^h(\mathbf{x}) = (1, 1, 1)^T, \ \forall \mathbf{x} \in K_i^{\tilde{p}}. \qquad (13)$$

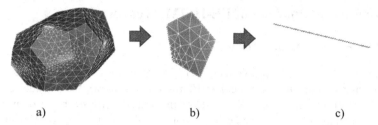

Fig. 4. The hierarchical system of meshes in a macroelement for a series of nested subproblems to construct the nonpolynomial shape functions: a – the tetrahedral mesh in the whole macroelement; b – the triangular mesh on the face (extracted from the tetrahedral mesh); c – the one-dimensional mesh on the edge (extracted from triangular meshes)

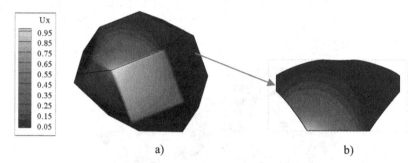

Fig. 5. Example of the x-component of the shape function on a polyhedron (a) and one of the sections (b)

The discrete variational formulation at the macrolevel has the form:

$$
\begin{cases}
\text{find } \mathbf{U}^H \in \mathbf{V}^H\left(\Pi^H(\Omega)\right) + \mathbf{U}_D(\Gamma_D) \text{ such as} \\
\displaystyle\int_\Omega \nabla \mathbf{U}^H : \mathbf{D}^H : \nabla \mathbf{v}^H d\Omega = \int_{\Gamma_N} \mathbf{G}_{Ng} \mathbf{v}^H(\mathbf{x}) d(\partial\Omega) + \int_\Omega \mathbf{F} \mathbf{v}^H d\Omega, \\
\forall \mathbf{v}^H(\mathbf{x}) \in \mathbf{V}^H\left(\Pi^H(\Omega)\right),
\end{cases}
\tag{14}
$$

where $\mathbf{D}^H$ is a homogenized material coefficient.

Thus, the solution of the problem (1)–(4) is represented at each point of the domain Ω as a decomposition over the corresponding shape functions:

$$
\mathbf{U}^H(\mathbf{x}) = \sum_{K_i^{\tilde{p}} \in \Pi^h(\Omega)} \sum_{p=1}^{\tilde{p}} \mathbf{q}_p^H \Psi_p^h(\mathbf{x}) = \sum_{K_i^{\tilde{p}} \in \Pi^h(\Omega)} \sum_{p=1}^{\tilde{p}} \mathbf{q}_p^H \sum_r \mathbf{q}_r^{p,h} \boldsymbol{\xi}_r^h(\mathbf{x}),
\tag{15}
$$

where $\mathbf{q}_p^H$ and $\mathbf{q}_r^{p,h}$ are the weights of degrees of freedom obtained from the solution of (12) and (14), respectively, $\boldsymbol{\xi}_r^h(\mathbf{x}) \in \mathbf{V}^h\left(K_i^{\tilde{p}}\right)$ are the basis functions from the space defined in (11).

3 Numerical Estimation of FE-HMM Accuracy

3.1 Homogeneous Computational Domain

Let the solution $\mathbf{U}^{FEM}$ of the problem (1)–(4) be obtained by the classical finite element method using a detailed mesh (more than 10 million tetrahedra). We consider $\mathbf{U}^{FEM}$ as an exact solution for all variants. To calculate the relative error of the z-component of the solution obtained by using the heterogeneous multiscale finite element method, we apply the L_2-norm as:

$$err = \frac{\left\| U_z^{FEM} - U_z^{FE-HMM} \right\|_{L_2}}{\left\| U_z^{FEM} \right\|_{L_2}} = \left(\frac{\int\limits_\Omega \left(U_z^{FEM} - U_z^{FE-HMM} \right)^2 d\Omega}{\int\limits_\Omega \left(U_z^{FEM} \right)^2 d\Omega} \right)^{1/2}. \tag{16}$$

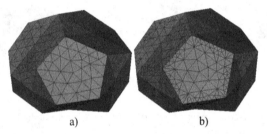

a)   b)

Fig. 6. Micromeshes of a macroelement with equal initial criteria for fineness of partitions: a – without adaptation, b – with additional sub-partitioning.

Table 1 shows a comparison of two approaches to the construction of meshes at the microlevel: without adaptation and with adaptation. In the variant with adaptation, additional subdivision of the tetrahedral mesh was performed on those edges of the polyhedron, where at the standard step of construction there were less than 5 nodes of the tetrahedral mesh (see Fig. 6).

As can be seen from the analysis of the results in the Table 1, the convergence of FE-HMM is hierarchical. The relative error of the solution decreases both by increasing the number of polyhedra at the macrolevel, and by splitting the mesh at the microlevel. However, in the variant with the micromesh without adaptation, there is a limit of the fineness at the macrolevel. Namely, on the mesh with 1290 polyhedra, the relative error is higher than on the meshes with fewer polyhedra for comparable fineness of the micromeshes. This is due to the fact that at a given criterion of micromeshes the necessary accuracy is not achieved when constructing the macrolevel shape functions even in a homogeneous medium. The introduction of an additional criterion of mesh adaptation along the edges allows to control this accuracy. This strategy significantly increases the total number of the degrees of freedom at the microlevel.

Table 1. Numerical characterization of computational meshes for a homogeneous sample and comparison of the two approaches to constructing meshes at the microlevel (N^h is the total number of tetrahedra at the microlevel)

Number of poly-hedrons	Micro-partitioning fineness criterion	Without adaptation		With adaptation	
		N^h	Relative error for U_z	N^h	Relative error for U_z
26	0.03	136 235	1.44E−02	135 650	1.48E−02
	0.02	490 329	1.42E−02	534 517	1.40E−02
	0.015	957 375	1.40E−02	1 131 719	1.40E−02
	0.01	4 106 514	1.40E−02	4 517 391	1.41E−02
76	0.03	142 146	5.74E−02	242 252	8.70E−03
	0.02	454 084	7.65E−03	549 335	7.88E−03
	0.015	1 194 948	7.38E−03	1 300 279	7.54E−03
	0.01	4 017 742	7.35E−03	4 237 923	7.63E−03
135	0.03	143 046	3.45E−01	431 854	6.82E−03
	0.02	522 917	4.37E−02	628 431	6.68E−03
	0.015	1 225 841	5.69E−03	1 275 817	5.94E−03
	0.01	4 551 570	5.46E−03	4 527 933	5.75E−03
217	0.03	155 259	4.81E−01	686 404	6.73E−03
	0.02	493 548	1.78E−01	746 568	6.27E−03
	0.015	1 271 845	5.61E−03	1 398 700	5.65E−03
	0.01	4 279 962	5.35E−03	4 336 714	5.11E−03
575	0.03	375 650	8.00E−01	2 282 913	5.76E−03
	0.02	534 636	4.93E−01	1 831 935	4.16E−03
	0.015	1 122 026	3.13E−01	2 517 785	3.98E−03

(*continued*)

Table 1. (*continued*)

Number of poly-hedrons	Micro-partitioning fineness criterion	Without adaptation		With adaptation	
		N^h	Relative error for U_z	N^h	Relative error for U_z
	0.01	4 290 063	3.68E−03	4 290 063	3.68E−03
1290	0.03	756 602	8.44E−01	5 156 084	1.33E−02
	0.02	769 786	8.33E−01	3 618 022	3.79E−03
	0.015	1 159 803	5.53E−01	4 396 185	4.00E−03
	0.01	3 884 906	3.16E−01	5 883 967	3.69E−03

3.2 Computational Domain with Inclusions

One of the basic problems of multiscale methods is to realize a technology that allows to work with unstructured media. In these media, it is impossible to construct a macromesh that its internal boundaries do not cross medium interfaces, and inclusions lie in some macroelement strictly inside. Polyhedral media allow us to reduce such intersections, since the shape of the macroelements can be varied over a wide range. However, it is not always realizable. For example, it is impossible for media with lenticular inclusions (Fig. 7a).

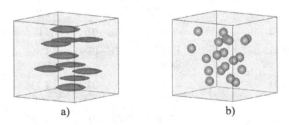

a)   b)

Fig. 7. Idealized macrostructure of permafrost rock.

To investigate the accuracy of the solution obtained in inhomogeneous objects, let us consider a two-component medium in the form of a cube with the edges of 1 m. The object consists of a matrix and inclusions of different configurations (Fig. 7). The matrix is sandstone (Young's modulus is 5 GPa, Poisson's ratio is 0.27) similar to the homogeneous sample from the point above. The inclusions are ice with Young's modulus 9.8 GPa and the Poisson's ratio 0.34. We assume the inclusions can be represented by the lenticular formations (two axises have fixed length 0.4 m, the third axis length varies randomly from 0.05 to 0.1 m) and the classical porous (diameter of spheres is 0.12 m).

For physically correct accounting of such peculiarities and preserving the continuity of the solution at the intersection boundaries, additional reduced subproblems (12) are solved to construct the macrolevel shape functions. The Table 2 shows the relative errors for the samples with lenticular inclusions (Fig. 7a) and spheres inclusions (Fig. 7b), for which the intersection of macroelements boundaries was also not controlled. The micromeshes were constructed with a conditional fineness criterion of "0.01" and with adaptation, since the most reliable results on a homogeneous sample were obtained at such parameters.

As in the case of the homogeneous sample (Table 1), the convergence with polyhedral mesh refinement is observed with a comparable number of tetrahedrons at the microlevel. However, 1290 polyhedra do not give a significant increase in accuracy, even when the number of microelements increases. The errors of the solutions for the sample with relatively large inclusions-lenses are comparable to the results for the samples with inclusions-spheres. Thus, our modification allows us to fully apply FE-HMM for media with inhomogeneities whose characteristic size exceeds the characteristic size of macroelements.

Table 2. Numerical characterization of the computational meshes and the relative errors of the solution for the sample with inclusions-lenses and inclusions-spheres (N^h is the total number of tetrahedrons at the microlevel)

Number of poly-hedrons	Sample with inclusions-lenses		Sample with spherical inclusions	
	N^h	Relative error for U_z	N^h	Relative error for U_z
26	4 502 546	2.01E−02	4 706 431	1.84E−02
76	4 353 217	1.48E−02	4 271 016	1.40E−02
135	4 511 366	1.12E−02	4 557 437	9.69E−03
217	4 385 528	1.09E−02	4 391 696	7.64E−03
575	4 519 899	9.63E−03	4 404 501	6.12E−03
1290	6 323 232	9.72E−03	6 138 601	5.38E−03

4   Parallelization of Computing Schemes

In this this paper, we will exclude the procedures for constructing hierarchical finite element meshes, as it is difficult to evaluate them in terms of the required resources, since a third-party software package SALOME is involved. For each macroelement the micromesh is also constructed independently.

In terms of implementation, the algorithm of the heterogeneous multiscale finite element method consists of the following steps:

1. A polyhedral macromesh is read and processed (e.g., a neighborhood relation is formed).

2. For each macroelements non-polynomial shape functions are independently constructed:
   a. the tetrahedral mesh of the macroelement is read and processed (including submeshes on each face and on each edge);
   b. the edges are analyzed and the traces of the shape function are formed along the one-dimensional boundary: if there are no homogeneities on the edge, the analytical relation is substituted, if there are homogeneities on the edge, the subproblems are solved;
   c. form traces of shape functions on quasi two-dimensional faces (in the general case the faces may not be planar) through the solution of a series of subproblems;
   d. systems of linear algebraic equations (SLAE) are formed and solved for each of the shape functions taking into account the corresponding traces on the boundaries.
3. Assembling the SLAU taking into account global boundary conditions and solving.
4. Analysis and output of results.

As it is not difficult to see, both algorithmic and software optimization are possible at each of the stages. However, in any case, the second stage generates a large number of additional subproblems since in a polyhedral medium, the number of shape functions coincides with the number of vertices which can be more than 30. Nevertheless, the processing on each of the macroelements in FE-HMM can be performed completely independently of the neighboring elements. Within the framework of this work a parallel version of the algorithm based on the OpenMP technology was realized. The Fig. 8 shows measurements of the solution time of the problem (1)–(4) for the homogeneous sample in the different variants of the meshes containing 26, 76, and 135 polyhedrons and preserving the fineness of micro-partitioning (about 1 million tetrahedrons).

The computational experiments were performed by using a computer with AMD Ryzen Threadripper PRO 3975WX 32-Cores (3.50 GHz) processor.

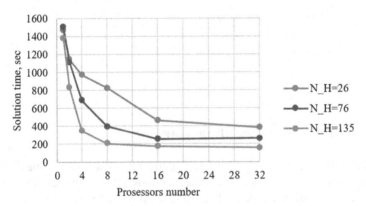

Fig. 8. Dependence of the problem solution time on the number of processors used.

5 Conclusions

A parallel modification of the heterogeneous multiscale finite element method using polyhedral supports is proposed for solving the problem of elastic deformation of a heterogeneous solid.

The developed algorithms and computational schemes are analyzed in terms of numerical convergence and requirements to computational resources. The "two-level" convergence specific to hierarchical finite element methods has been shown experimentally. The solution accuracy can be controlled by varying the discretization at the macrolevel as well as at the microlevel. In this case, to achieve the best accuracy of the solution, it is necessary to maintain a balance in mesh refinement both in the construction of nonpolynomial shape functions at the micro- and macrolevels. The proposed automatic adaptation of the mesh at the microlevel near the edges of the polyhedron allows to obtain a more accurate solution (up to two orders of magnitude) at comparable sizes of discretizations. The approach in parallel realization of computational schemes used within the framework of this stage allowed to reduce the solution time up to 10 times on 32 threads. Also, in the future, it is planned to modify the technology of load distribution over processor threads to increase efficiency.

Acknowledgements. This work was carried out with the financial support of the RSF (project No. 22-71-10037).

References

1. Zienkiewicz, O.C., Bahrani, A.K., Arlett, P.L.: Numerical solution of 3-dimensional field problems. Proc. IET **16**, 367–369 (1968)
2. Weinan, E., Ming, P., Zhang, P.: Analysis of the heterogeneous multiscale method for ellpiptic homogenization problems. J. Am. Math. Soc. **18**, 121–156 (2003)
3. Abdulle, A., Weinan, E., Engquist, B., Vanden-Eijnden, E.: The heterogeneous multiscale methods. Acta Numer. **21**, 1–87 (2012)
4. Epov, M.I., Shurina, E.P., Itkina, N.B., Kutishcheva, A.Y., Markov, S.I.: Finite element modeling of a multi-physics poro-elastic problem in multiscale media. J. Comput. Appl. Math. **352**, 1–22 (2019)
5. Eidel, B., Fischer, A.: The heterogeneous multiscale finite element method for the homogenization of linear elastic solids and a comparison with the FE2 method. Comput. Methods Appl. Mech. Eng. **329**(1), 332–368 (2017)
6. Aurenhammer, F.: Voronoi diagrams - a survey of a fundamental geometric data structure. ACM Comput. Surv. **23**(3), 345–405 (1991)
7. Frey, P.J., George, P.L.: Mesh Generation - Application to Finite Elements, p. 848. Wiley, London (2008)
8. Barber, C.B., Dobkin, D.P., Huhdanpapp, H.T.: The Quickhull algorithm for convex hulls. ACM Trans. Math. Softw. **22**(4), 469–483 (1996)
9. Rycroft, C.H.: Voro++: a three-dimensional Voronoi cell library in C++. Chaos **19**, 041111 (2009)
10. Yan, D.-M., Wang, W., Lévy, B., Liu, Y.: Efficient computation of 3D clipped Voronoi diagram. In: Mourrain, B., Schaefer, S., Xu, G. (eds.) GMP 2010. LNCS, vol. 6130, pp. 269–282. Springer, Heidelberg (2010). https://doi.org/10.1007/978-3-642-13411-1_18

11. Ebeida, M.S., Mitchell, S.A.: Uniform random Voronoi meshes. In: Quadros, W.R. (ed.) Proceedings of the 20th International Meshing Roundtable, pp. 273-290. Springer, Heidelberg (2011). https://doi.org/10.1007/978-3-642-24734-7_15
12. Garimella, R.V., Kim, J., Berndt, M.: Polyhedral mesh generation and optimization for non-manifold domains. In: Proceedings of the 22nd International Meshing Roundtable, pp. 313–330. Springer, Cham (2013). https://doi.org/10.1007/978-3-319-02335-9_18
13. Shurina, E.P., Epov, M.I., Kutishcheva, A.Y.: Numerical simulation of the percolation threshold of the electric resistivity. Comput. Technol. **22**(3), 3–15 (2017)

Parallelization of Finite-Volume Numerical Methods of Computational Fluid Dynamics by Means of Shared Memory Computing Systems

Anastasia Gulicheva[(✉)]

MIREA – Russian Technological University, Vernadsky Avenue 78, 119454 Moscow, Russia
gulicheva@mirea.ru

Abstract. The article considers the problems of parallelization of finite-volume numerical methods concerning the calculation of the streaming flow of conservative quantities through the boundaries of computational cells for systems operating with shared memory. Three approaches to solving this problem were analyzed, for which a theoretical justification was carried out, implementation, testing, analysis of the performance and scalability of the proposed methods were performed.

Keywords: Numerical Methods · Finite Volume · Parallelization · Shared-Memory Systems · Flows of Conserved Quantities · Computational Grids

1 Introduction

At present, there is a tendency in the world to develop supercomputer cluster systems, the computing nodes of which consist of a large number of cores and are capable of performing calculations with shared memory using a huge number of threads. Efficient use of such computing nodes becomes a priority. When performing calculations using finite-volume numerical methods, calculations require performing two phases of calculations at each iteration of calculations. At the first phase, within each computational cell, independent of the remaining cells, local physical quantities are recalculated (this phase can be ideally parallelized due to the independence of the calculations). During the second phase, the computational cells exchange information with each other by following streams of conservative quantities between them. At this phase, conflicts may arise over access to the data of computational cells, and the analysis of effective methods for parallelizing calculations at this phase is the subject of this work.

The main objective of this article is to implement computation parallelization using the graph coloring technique for conflicts arising from the streaming flow of conservative quantities of unstructured surface computational grids and for ordinary regular volumetric grids, the cells of which are rectangular parallelepipeds.

In the parallel calculation of physical quantities and the flow of their streams between cells on a shared memory computer, we encountered the following problem: in the parallel calculation of the flow of a quantity flow through two boundaries incident to the same cell, there is a conflict in access to the data of this cell. Therefore, for a correct calculation, such conflicting flows should be calculated strictly sequentially. To solve this problem, several methods of parallel computation of streams of conserved quantities were implemented: using OpenMP, using intermediate storage of data at cell boundaries, and using conflict graph coloring.

2 Problem Statement

In the calculations performed within the cells, physical quantities of two types are used (for example, the problem of gas dynamics): *primitive* (which characterize the state of matter) and *conservative* (for which conservation laws are satisfied).

$\vec{W}$ *vector of conservative variables* $\left(= [\rho, \ \rho u, \ \rho v, \ \rho w, \ \rho E]^T\right)$,

$\vec{W}_p$ *vector of conservative variables* $\left(= [\rho, \ \rho u, \ \rho v, \ \rho w, \ \rho E]^T\right)$.

Therefore, we obtain a system characterized by vectors of conservative variables $\vec{W}$, flow vectors $\vec{F}_c$ and $\vec{F}_v$, as well as original member $\vec{Q}$ extended equations of the form $(N-1)$ [1]. The vector of the conservative variables now looks like:

$$\vec{W} = \begin{bmatrix} \rho \\ \rho u \\ \rho v \\ \rho w \\ \rho E \\ \rho Y_1 \\ \vdots \\ \rho Y_{N-1} \end{bmatrix}.$$

The state of the substance in the cell is fully characterized by both the vector of primitive variables and the vector of conservative variables, and each can be obtained from the other using the corresponding transformation.

Each computational cell is adjacent to other computational cells, adjacent cells touch through a common boundary (for volumetric meshes, the boundary is a flat face, as in Fig. 1(b), for surface meshes, the cell edges are the boundary, as in Fig. 1(a)) through the common boundaries there is an overflow of flows expressed in conservative values.

In Fig. 1, D – primitive quantities, U – conservative quantities, and F – flows.

The general scheme of calculation and substance exchange for one cell in a single iteration can be outlined as follows:

1. calculation of primitive values D in the calculation cell;
2. obtaining a vector of conservative values from the vector of primitive values, that is, $D \rightarrow U$;

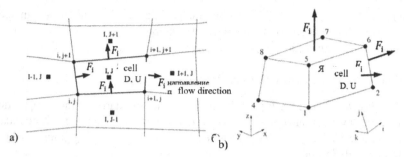

Fig. 1. Examples of cells in one-dimensional (a) and two-dimensional (b) grids.

3. for each boundary of the considered cell, the matter fluxes F_i are calculated, expressed in conservative values;
4. recalculation of the vector of conservative values taking into account flows through the boundaries $U = U + F_i S_i \Delta t$ for all boundaries;
5. obtaining a vector of primitive values from the vector of conservative values, i.e. $U \rightarrow D$.

When all flows from step 4 are calculated simultaneously, a conflict arises over access to the vector of conservative values of the cell. Specifically, all the streams that pass through boundaries incident to the same cell are involved in pairwise conflicts [8].

3 Parallelization of Computing Flows of Conserved Values

When using multi-threaded machines, there is a desire to perform thread recalculation in parallel mode. To ensure this, it is necessary to resolve conflicts on access to conservative values when they are corrected due to the flow of matter flows, and this can be resolved by implementing the following methods:

1. Using the OpenMP critical sections mechanism. Cons of the approach: the need to use critical sections, which can be problematic for some architectures (for example, for Elbrus).
2. Saving flows first at all boundaries, and then a one-time recalculation for each cell. To implement this mechanism, it is also necessary to store information about all boundaries that are input (through which flows enter) and output (through which flows exit) for the cell. The sequence of actions in this case:
   a. bypass all borders between cells, calculate the flows on them and save them on the borders;
   b. bypass all cells of the computational grid to the vector of conservative values, add the corresponding fluxes from the input boundaries (corrected for area and time step) and subtract the fluxes from the output boundaries.

   Cons of the approach: the need for each cell boundary to store information about whether this boundary is an input or an output; the need to maintain conservative values of flows at the boundaries.

3. Dividing the set of boundaries between the calculated cells into subsets in such a way that each subset does not contain conflicting boundaries. Cons of the approach: additional steps are needed to split the set of boundaries into subsets without conflicts.

3.1 Reduction of the Parallelization Problem

The graph is entered. The vertices of the graph are the boundaries of the computational grid. Two graph vertices are connected by an edge if the corresponding mesh boundaries conflict. Next, we solve the problem of vertex coloring of the resulting graph. To do this, we introduce a definition.

Definition 1. Let G be an undirected graph. A vertex coloring of a graph G is a mapping $f: (G) \to \mathbb{N}$ such that $(v) \neq (v)$ for all $\{v, w\} \in (G)$.

The numbers $f(v)$ are called the colors of the vertices v. In other words, sets of vertices with the same color (value f) must be independent sets or matchings, respectively. We are interested in using as few colors as possible [7].

The value of the optimum in the vertex coloring problem (that is, the minimum required number of colors) is called the chromatic number of the graph [11].

Theorem 1. Let G be an undirected graph with maximum degree k. Then there is a vertex coloring of the graph G, where no more than $k + 1$ colors, and this coloring can be found in linear time.

Theorem 2 (Brooks). Let $G = (V,)$ be a connected graph that is not a complete graph in which the greatest degree of a vertex is $d \geq 3$. Then the graph can be colored with d colors.

In this article, we use a greedy algorithm. Using it, the vertices $v_1, \ldots, v_n$ are ordered and sequentially assigned to the vertex v_i the smallest available color that was not used to color the neighbors vi among $v_1, \ldots, v_{i-1}$, or adds a new one. The quality of the resulting coloring depends on the chosen order.

In this work, we considered two variants of the greedy algorithm. The difference lies in the fact that in the first case, any arbitrary vertex is selected and assigned the first color, while in the second case, the first color is already assigned to all vertices of the graph. Subsequently, the greedy algorithm starts working in both cases.

3.2 Application of Graph Coloring in Parallelization

Let's declare a graph. The vertices of the graph are the boundaries of the computational grid. Two graph vertices are connected by an edge if the corresponding mesh boundaries conflict. Next, we solve the problem of vertex coloring of the resulting graph.

In this article, we used graph (Fig. 2) to show the results.

In Figs. 2 and 3 we see that the maximum degree of a vertex is four, which means that such a graph can be colored in 4 colors according to theorem 2, this ensures planarity [9].

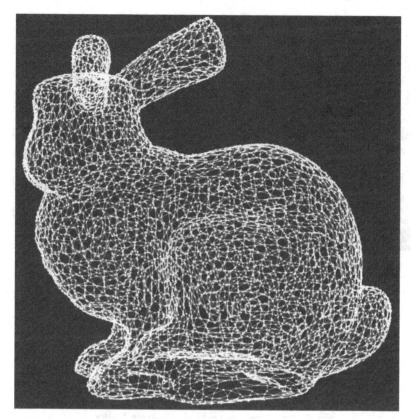

Fig. 2. An example of an unstructured surface mesh "Rabbit".

3.3 Optimized Greedy Coloring Method

Accordingly, the greedy coloring algorithm produces a result of 5 colors. Here, the first random vertex is selected, which is assigned the value of the first color, and then all subsequent vertices are colored in such a way that the color does not match the already colored vertices with which this vertex has a common edge. The color is selected from the already used colors, but if it is impossible to use them, a new color is added.

But we have developed a more optimal method, and when we run it, the result is: 4 colors. Now we describe the coloring algorithm in 4 colors [3, 10].

1. Since the degree of each vertex of the graph is at most 4, it is possible to paint in 5 colors in a greedy way. After that, we will try to recolor all the vertices painted in color 5.
2. Consider a vertex with color 5. If it cannot be recolored, then all its 4 neighbors are colored 1, 2, 3, 4 (Fig. 4).
3. If it is possible to recolor at least one neighbor in a color different from 5, then the vertex under consideration can then be recolored. This means that in the worst case, no neighbor can also be repainted, which means that the neighborhood of the considered vertex has the form, as shown in Fig. 5.

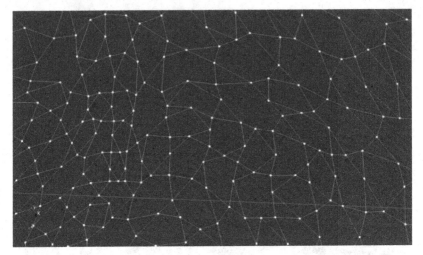

Fig. 3. An example of a close-up view of an unstructured surface mesh "Rabbit".

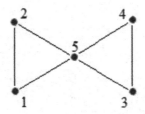

Fig. 4. An example of a case of coloring in 5 colors.

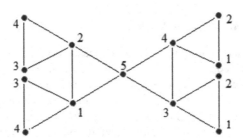

Fig. 5. Worst case coloring in 5 colors.

4. It follows from the appearance of the neighborhood that color 5 can migrate in any direction, for example, exchanging places with 4, as shown in Fig. 6.
5. With the help of the described color 5 migration mechanism, it is possible to drive all such colors to the graph boundary (where the degree of vertices is less than 4) and recolor in a greedy way. Since we are considering the worst-case scenario, we will assume that the graph has no boundaries.

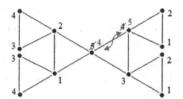

Fig. 6. Color replacement example.

6. If there is more than one vertex of color 5 in the graph, then using the migration mechanism, we bring them closer to each other in the following configuration, as shown in Fig. 7.

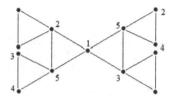

Fig. 7. The result of the migration.

After approaching two vertices of color 5 at a distance of 2 to each other, in this case we can recolor them to color 1, and recolor one to 5, thus reducing the number of vertices of color 5. Since we are considering the worst case, we will assume that we have one vertex of color 5 left in the graph.

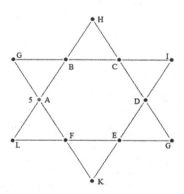

Fig. 8. An example of an even length loop.

7. Let's assume that there is only one vertex of color 5 left in the graph without a border, and no recoloring of this vertex can be done with any actions to migrate this vertex. Consider one of the larger cycles that this vertex (A) enters, as shown in Fig. 8.

Let this cycle be of even length. For definiteness, let's draw a cycle of length 6, as shown in the figure. Since the vertex A cannot be recolored, without loss of generality we will assume that the vertices G and B are colored in colors 1 and 2 (which of them is colored in color 1 and which in color 2 does not matter), and the vertices L and F are colored in colors 3 and 4. We agreed to assume that after the migration of color 5 to vertex B, this vertex B cannot be recolored. Since, after the migration of color 5 to vertex B, the vertices AG will be colored in colors 12, then the edge HC must be colored in colors 34 (we denote HC ~ 34).

By analogous reasoning, we get DI ~ 12, EG ~ 34, FK ~ 12. However, FK ~ 12 contradicts LF ~ 34. This means that at some point during the migration of color 5 along an even cycle, we could recolor the vertex. This can be seen in Fig. 9. Let us similarly consider the case when the large cycle is of odd length. To be specific, let's draw a cycle of length 5 as shown in Fig. 10 below.

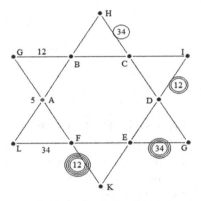

Fig. 9. An example of a contradiction in a cycle of even length.

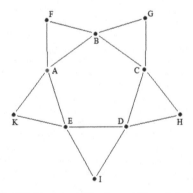

Fig. 10. An example of an odd length loop.

And once again, we start reasoning from the fact that since vertex A cannot be recolored. Without loss of generality, let's assume that vertices F and B are colored with colors 1 and 2. It doesn't matter which one is colored with color 1 and which one with color 2. Similarly, vertices K and E are colored with colors 3 and 4. We agreed to assume that after the migration of color 5 to vertex B, this vertex B cannot be recolored. Since after the migration of color 5 to vertex B the vertices AF will be colored in colors 12, then the edge GC must be colored in colors 34 (we denote GC ~ 34). By similar reasoning, we get DH ~ 12, EI ~ 34, AK ~ 12, which contradicts the fact that KE are colored in colors 3 and 4. This means that at some point during the migration of color 5 along an odd cycle, we could recolor top. The result is in Fig. 11.

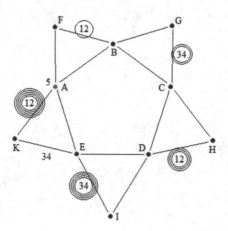

Fig. 11. An example of a contradiction in a cycle of odd length.

3.4 Verification of the Algorithm

The result of the algorithms described above is shown in Fig. 12, where you can see that the presented graph can be colored not in 5 colors, but in 4. The vertices that have two colors are those vertices that have been repainted, they have an outer rim - this is the color that the greedy algorithm produced, the central color is the color in which they can be recolored to implement a more optimal coloring.

4 Discussion

To measure the scalability of computations on an unstructured surface computational grid, we used a test surface of a streamlined three-dimensional body containing about 105 nodes and 105 cells. In the cells, calculations were performed related to modeling the flow of a liquid film, solving the heat balance equations on the surface, as well as restructuring and smoothing the surface (Fig. 13).

The calculations were performed on one computing node. The main goal of the launches was to measure the strong scalability of computing using a different number of threads within the computing node [2, 5, 6].

That is, the same surface was used for all launches (which was provided by a different number of threads).

Fig. 12. The result of optimizing the greedy algorithm on the "Rabbit" graph.

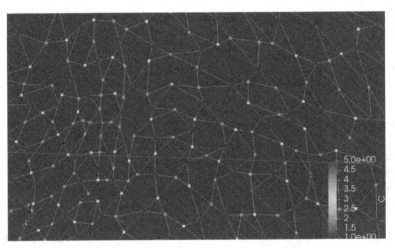

Fig. 13. The result of the optimization of the greedy algorithm on the "Rabbit" graph near.

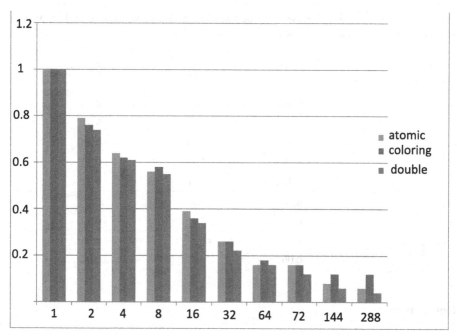

Fig. 14. Efficiency of computing scaling on a single computing node based on the Intel Xeon Phi KNL microprocessor microprocessor with an increase in the number of threads.

Figure 14 shows a diagram of the efficiency of computing scaling for various methods of parallelizing thread computations with an increase in the number of threads within a single node. At the same time, the value of

$$s(i) = t(1)/i(1),$$

where $t(1)$ is the reference time, was taken as acceleration at the number of nodes equal to i. In this case, the computational scaling efficiency is understood as the value of

$$e(i) = s(i)/i.$$

Its physical meaning is as follows. It can be assumed that with an ideal parallelization of calculations, with an increase in the number of threads by a factor of n, the execution time decreases exactly by a factor of n. Thus, in the case of ideal parallelization, $(i) = i$, and $(i) = 1$. The scaling efficiency of computations is a convenient indicator of the quality of creating an executable parallel code and comparing different computing systems with each other [4]. Note that superlinear scalability is quite possible (when the value of (i) becomes greater than 1 (Table 1).

Thus, we can conclude that with an increase in the number of computational threads, the most efficient way is to divide edges (borders between computational cells) into independent sets (graph coloring) and process them in turn.

Table 1. Calculation scaling efficiency indicators for various methods of thread computing parallelization with an increase in the number of computing nodes used.

Number of threads	atomic	coloring	double
1	1	1	1
2	0.79	0.76	0.74
4	0.64	0.62	0.61
8	0.56	0.58	0.55
16	0.39	0.36	0.34
32	0.26	0.26	0.22
64	0.16	0.18	0.16
72	0.16	0.16	0.12
144	0.08	0.12	0.06
288	0.06	0.12	0.04

References

1. Vorobyov, E.S., Vorobiev, V.E.: Numerical methods and mathematical modeling. Fundamentals of numerical methods and techniques for building mathematical models based on them and these solutions in various packages, Kazan (2017)
2. Rybakov, A.A.: Distribution of the computational load between the nodes of a heterogeneous computing cluster. Progr. Products Syst. Algorithms **1**, 1–7 (2018)
3. Golovchenko, E.: Overview of Graph Decomposition Algorithms. Preprint IPM № 002 (IPM named after M. V. Keldysh) (2020)
4. Rybakov, A.: Internal representation and mechanism of interprocess exchange for a block grid for supercomputing. Software products and systems. Algorithmiya **8**(1), 121–134 (2017)
5. Software Developer's Guide for Intel 64 and IA-32 Architectures (Intel Corp.), Consolidated Volumes: 1, 2A, 2B, 2C, 2D, 3A, 3B, 3C, 3D, and 4 (2019)
6. Jeffers, J., Reinders, J., and Sodani, A.: High-Performance Programming of Intel Xeon Phi Processors, Knights Landing Edition. Morgan Kaufmann, Burlington (2016)
7. Kuhn, F.: Weak graph colorings: distributed algorithms and applications. In: Proceedings of the 21st Symposium on Parallelism in Algorithms and Architectures, pp. 138–144 (2009)
8. Blazek, J.: Computational Fluid Dynamics: Principles and Applications. Elsevier, Amsterdam (2001)
9. Molloy, M., Reed, B.: Graph colouring and the probabilistic method. Springer, Heidelberg (2002). https://doi.org/10.1007/978-3-642-04016-0
10. Korte, B., Vygen, J.: Combinatorial Optimization Theory and Algorithms. Springer, Heidelberg (2015). https://doi.org/10.1007/978-3-662-56039-6
11. Anferov, M.A.: Algorithm for searching subcritical paths on network graphs. Russian Technol. J. **11**(1), 60–69 (2023)

Computing Technologies in Data Analysis and Decision Making

Development and Comparative Study of Data Compression Methods Used in Technical Monitoring Systems

Alexey G. Yakunin$^{(\boxtimes)}$

Polzunov Altai State Technical University, Prospect Lenina 46, 656038 Barnaul, Russia
almpas@list.ru

Abstract. The article presents the results of the development and research of methods for compactifying the transmission and storage of data generated by technical monitoring systems. The research was carried out using the information and measurement system of the Altai State Technical University. The system is designed to monitor the air temperature in the premises of the university campus and surrounding area, and consumption of resources accounting (heating, hot and cold water). Developed and modified lossy and lossless data compression methods and their experimental studies are considered, including online compression methods during data transfer to the server. Database structures designed to store information collected during the monitoring process are also described. The study of the compression efficiency of the developed algorithms was carried out mainly on temperature monitoring data. This was due to the fact that control and monitoring of temperature processes is most widespread in various technical process control systems, heat metering systems both on the consumer side (housing and communal services), and on the side of the heat supply organization, and during meteorological observations. The results of the studies demonstrate that in most cases, lossless compression methods allow to compress the information received from the monitoring system by more than 10 times, and lossy compression methods – even more, without losing any pragmatic value of the original information.

Keywords: Data Compactification · Lossless Compression · Lossy Compression · Technical Monitoring · Data Redundancy Elimination

1 Introduction

Monitoring systems are widely used in various spheres of human activity, including, but not limited to, financial, industrial, scientific. Technical monitoring systems represent a set of software and hardware, sensors and other equipment that makes it possible to systematically observe and record physical quantities that describe temporary changes the states of any objects and processes occurring in them. These can be technical, biological and natural systems and objects, buildings and structures, and environmental objects located inside or outside: both liquid, solid and gaseous [1].

© The Author(s), under exclusive license to Springer Nature Switzerland AG 2024
V. Jordan et al. (Eds.): HPCST 2023, CCIS 1986, pp. 119–131, 2024.
https://doi.org/10.1007/978-3-031-51057-1_9

An important function of monitoring systems is recording and store a time sequences of recorded values of observed parameters. The duration and volume of such records depend on the intended purpose of the monitoring system, the main of which are control, management and observation.

For monitoring systems which purpose is control and management, the main function is to check the compliance of the parameters of the monitored object with the specified values and the nature of their change over time and formation of control actions and, if such compliance is violated, informing the personnel servicing the system. Therefore, such systems generally do not require long-term recording of observed results, and sometimes do not require recording them at all.

If the purpose of the monitoring system is observation, then its main function is to register the observed values and store them for the preparation and generation any reporting documents in the future or for analysis and processing during research. An example of such research would be studies aimed at identifying patterns of global climate change or changes in solar activity, which require at least decades of data. Such long-term recording can also help in investigation of emergency situations that may occur in a managed system in order to determine the causes that led to them. Without it impossible to imagine the operation of accounting systems for the supply of thermal or electric energy at thermal and electric power plants, when, according to the regulations, such information must be stored for at least an year. Similar terms are provided for the logs of registration of the temperature regime of refrigeration equipment. This article is devoted to consideration of issues of optimizing methods for processing, transmitting and storing large volumes of data generated by technical monitoring systems.

2 Problem Description

Modern technical systems are characterized by a large number of controlled parameters, and the value of some of them can change very quickly. Often it may also be necessary to transfer a large flow of measuring information between different subsystems of control systems via telemetry channels. And, as already noted earlier, in a number of cases there is a need to store the measured results for a sufficiently long time interval. Hence the need to solve issues of compactification of both transmitted and stored data generated in such systems becomes obvious. To date, a huge number of works and information resources are devoted to data compression issues, for example, [2–4], most of which are devoted to multimedia information compression. At the same time, works began to appear devoted specifically to the compression of data generated by sensors of information-measuring systems [5–7], and the use of non-traditional compression algorithms for other types of data, not images and sound [8–10].

The aim of this article is to present particular research results related to the development of effective high-performance methods of information compression and algorithms that implement them. The data used in this study was taken over a period of more than 12 years from the information and measurement system of the Altai State Technical University. A detailed description of this system is given in [11, 12].

Initially, the main purpose of this system was to collect data on the consumption of energy resources by the university, but later, after the replacement of metering devices, the focus was shifted to monitoring weather data and air temperature processes. And, for the reasons previously stated, it was temperature monitoring data that was used in most experiments studying compression algorithms. Specifically, these were observational data on air temperature in rooms and near buildings located on the university campus.

3 Methods Based on Data Normalization and Structuring

When solving technical monitoring problems, the volume of stored data can be optimized by normalizing and structuring it, taking into account the type of measured values and the dynamics of its change over time. So, in the monitoring system described in [11–13] for data storage, a database optimized from these points of view was used, the structure of which is shown in Fig. 1. This database was implemented in the DBMS (database management system) MySQL.

The main database table is the "chnl" table, which stores information about all measuring channels of the system. In relation to the topic of this work, the fields "id", "calc", "table", "clock", "mult" and "shift" are involved in it. The "id" field specifies the number of the channel from which data is received, and the "table" field indicates one of the tables "ds", "di", "dh", "d", " dd", "dm" (respectively, with a multiple of a second, minute, hour, daily or monthly interval). In these tables, the "chnl" field corresponds to the channel number specified in the "id" field of the "chnl" table, and the "val" and "clc" fields store the normalized values of the initial or processed measurement results and the reading numbers n_i.

In the system, the minimum indivisible time interval is the interval of the base cycle $\Delta_t = 30s$. Studies have shown that there was no need for more frequent temporal discretization for the physical quantities recorded in the system. All other time intervals used for sampling measurements carried out in different channels are multiples of 30 s and the multiplicity factor k_t, which specifies the number of base cycles between adjacent readings of the controlled parameter, is stored in the "clock" field. Therefore, in the tables "ds", "di", "dh", "dd", "dm", the real time t_i, at which the countdown was made, is found from the expression $t_i = t_o + k_t \cdot n_i \cdot \Delta_t$, where t_0 is the date and time of the database initialization specified in the initialization file. Such a solution made it possible to significantly reduce the space allocated in the database for storing information about the time of obtaining readings [14]. If the "clk" field stored the absolute time of the corresponding measurement, then its size in the MySQL DBMS would be 8 bytes, whereas, as can be seen from Table 1, with the proposed option for storing time readings, a field length of 2–3 bytes is more than sufficient for this.

It should be emphasized that with this approach, the length of the "clk" field should be set very carefully. In particular, in the information system described in [11–13] at the time of its initialization in 2009, the sizes of the fields highlighted in the Table 1 in gray were initially selected. As a result, in some tables, namely, in tables "ds" and "di", after 5 and 10 years, respectively, it became necessary to increase the field size by 1 byte (in Fig. 1 for tables, the initially selected field types are shown). Moreover, when working with the database, it turned out that it is better to set the t_0 value not as a constant, the same for all channels, but to indicate this value for each channel separately by entering an additional field in the "chnl" table, since during system operation monitoring channels may come into effect at different times.

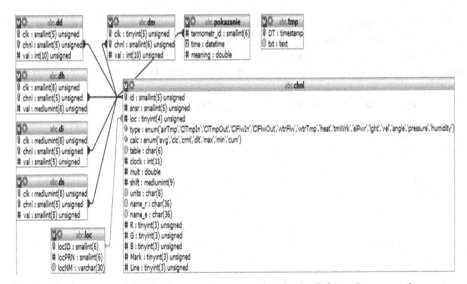

Fig. 1. The structure of the database used to store data in the information-measuring system described in [11, 12].

Table 1. Dependence of the maximum possible storage time (specified in years), on the data-sampling interval and the size of the "clk" field specified in bytes.

Number	*Data observation interval*									
of bytes	*1c*	*10s*	*30s*	*1min*	*5min*	*10min*	*30min*	*1 h*	*1 day*	*1 month*
1									0.7	21
2						1.3	4	7.5	180	
3	0.53	5.3	10.6	32	160	133				
4	136									

The "calc" field in the "chnl" table indicates the nature of the processing of the measurement information coming from the channel, the results of which will be stored in the system. Such operations can be averaging data, selecting the maximum, minimum or accumulated value, or refusing to perform any operations and saving the data in the original form (when the value of the "calc" field is set to "crnt"). Depending on the type of data processing and the sampling interval, after processing this data is written to the "val" field of one of the tables "ds", "di", "dh", "dd" or "dm" in accordance with the value of the field "table" of the "chnl" table. In this system, the results of temperature measurements and wind speed were recorded in the table "ds", information on pressure and humidity was recorded in the "di" table, readings of heat and water meters were recorded in the "dh" table, and the accumulated meter readings per day and per month stored in the "dd" and "dm" tables, respectively.

In any case, a mandatory operation before recording is the operation of data normalization, since the data is registered in the system from the measuring channels as real numbers that correspond to the actual values of the measured parameter in accordance with the measurement units specified in the "units" field. These numbers are converted into normalized integer values using the formula $y_i = round(k_x \cdot x_i + y_0)$, where x_i is the data from the channel, k_x is the multiplicative scaling factor of the initial value of the parameter, y_0 is offset, and *round* is a rounding function to an integer value. The values of the factors are selected so that the minimum value of the measured parameter corresponds to a zero value y_i, and the maximum value corresponds to a value close to 2^{8n}, where n is the number of bytes allocated for storing the measurement results of the observed parameter. As a result of such normalization, data storage will require a field with a length of only 2–3 bytes instead of 8, since the measurement errors are extremely rarely less than 0.1–0.001%, therefore, with a minimum of computational operations, data can be compressed by 2–4 times.

Another feature of the proposed structure is an ability to store the same parameter in several tables at once, depending on the nature of the processing. Thus, in the developed database, the current water meter readings are stored in the "ds" table, hourly flow rates are accumulated in the "dh" table, daily flow rates are accumulated in the "dd" table, and monthly costs are accumulated in the "dm" table. Although the volume of stored data in this case increases slightly, it significantly increases an efficiency of obtaining the necessary information. In future, further improvements to the database structure are expected to combine high compression rates and fast data access. To achieve this, the methods discussed below are supposed to be applied to maximize the compression of archived data from past periods (more than a year) and use less resource-intensive compression methods to access operational data of current periods. This approach is quite justified, since the data of past years are less in demand and are more often used for research work, when the time of access to the data is not so important, since it is still be much less than the time required to conduct the study itself.

4 Compression Methods Based on Difference Schemes and RLE-Algorithms

A further increase in the degree of compression can be achieved through the use of difference schemes [15]. The principle of such compression is the following. Instead of the original signal registered by sensors, its deviations from the average value, taken over a certain time interval (in this case, a day), are stored (See Fig. 2).

As follows from Fig. 2, deviations from the average value are 2–4 times less than the values of the original signal, which makes it possible to reduce the dimension of the bit grid required for data storage by 1–2 bits. Deviations can be reduced even more significantly by reducing the averaging interval. It should be noted, however, that in addition to deviations, it is also necessary to store the average values themselves, which reduces the degree of compression. Also, if there is a significant fluctuation in the observations or an up or down trend, as is the case in Fig. 2, the difference values might quickly become too large and will not differ much from the initial reference value. Therefore, the criterion when the difference values reach a given threshold value that will be the main reason for limiting the number of elements of the time series. Consequently, the base difference method is not effective enough, and in practice, schemes are more often

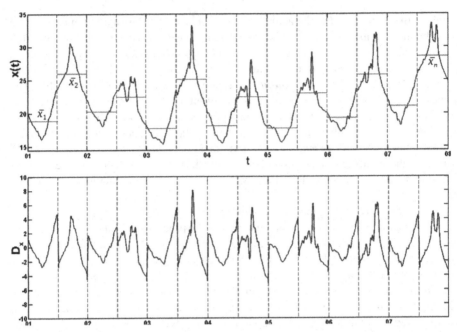

Fig. 2. Time diagrams of cyclic change in daily street temperature $x(t)$ and difference signals $D(t)$ between these changes and its average daily values $\bar{x}_i$.

used when not deviations from one control point (be it an average value or some kind of absolute reading) are stored, but the difference in the values of neighboring readings.

As an example, in Fig. 3, various typical graphs of temperature changes on the street and at various points in the room are presented. Temperature measurement was carried out using DS18B20 sensors with a temperature resolution of 0.0625 °C. As can be seen from graphs, even with the sharpest temperature fluctuations, the difference in the values of adjacent readings does not exceed 32 quantization levels (Fig. 3b, the last right temperature spike). Therefore, it is enough to allocate not 10, but only 5, bits for one reading almost to halve the amount of stored data with virtually no decrease in the speed of their subsequent sampling.

Therefore, it is enough to allocate not 10, but only 5 bits for one reading in order almost to halve the amount of stored data with virtually no reduction in the speed of their subsequent retrieval.

To increase reliability and reduce data recovery time, the sequence of difference values must be periodically interrupted and started again so that the numerical series forming a series of readings has a limited length. Then the loss or error of a single initial absolute value will not lead to a complete loss of information about all the series of readings included in the observation of the controlled physical quantity, and in order to restore the reading values for the last observation period, it would not be necessary to perform a complete recalculation, starting from the very first value of the time series.

The structure of the database intended for storing the registered and packed readings in this way is shown in Fig. 4. The entire time series formed by the normalized values

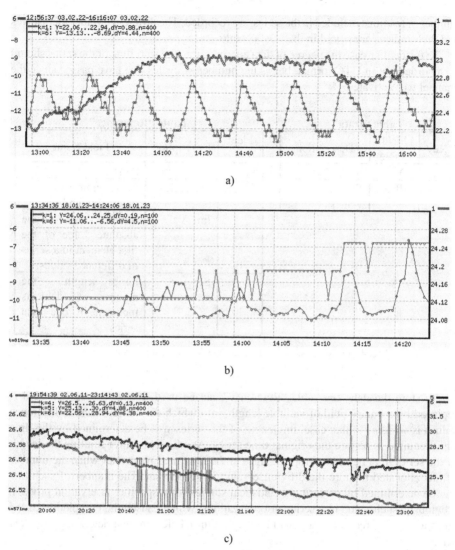

Fig. 3. Time series of outdoor temperature (blue line, channel 6), temperature at various points in the room (red line, channel 1 - temperature at the ceiling and brown line, channel 4 - at the floor) and in the attic (channel 5, black line). a) - an electric heater is working, creating regular temperature oscillations in the room, b) strong gusts of wind, leading to temperature jumps up to 2 °C/s due to the presence of micro-turbulences in the air, c) - calm, gradual cooling by night. (Color figure online)

of the measured value is divided into separate series. In the "Reference" table for each series, the measurement time and the absolute value of the first reading included in the series, as well as the index corresponding to the first element of series and the total number of readings included in the series, are recorded. The difference values themselves and their indices are stored in the "Data" table.

To further increase the degree of data compression, the start time of the series should be recorded not in UTC format, but as the number of cycles that have passed since the channel was activated.

<table>
<tr><td colspan="3" align="center">Table "Data"</td><td colspan="3" align="center">Table "Reference"</td></tr>
<tr><td>Key</td><td>Field Name</td><td>Description</td><td>Field Name</td><td>Description</td><td>Field Name</td></tr>
<tr><td>PK</td><td>ID</td><td>Series Index</td><td>PK</td><td>Start_time</td><td>Series start time</td></tr>
<tr><td></td><td>Offsets</td><td>Difference values</td><td>FK</td><td>Series_ID</td><td>Series number (or index, foreign key)</td></tr>
<tr><td></td><td></td><td></td><td></td><td>Init_Val</td><td>Initial Value - the absolute value of the first element in a series of difference values</td></tr>
<tr><td></td><td></td><td></td><td></td><td>Ser_Len</td><td>Series length - the number of elements included in one data series</td></tr>
</table>

Fig. 4. Database structure for storing series of difference readings

To store the data, one table "Reference" can be used, if the "Series_ID" field is replaced by a binary field of variable length, into which the entire series of difference values is written. In this case, the length of the series will be determined by the time limit required to restore the original time series. Another reason for limiting the length of the series may be large abrupt changes in the measured parameter when the number of bits allocated to store the difference value is insufficient. The number of such bits can be determined by performing statistical studies for a specific observation process. Obviously, such value in a series of observations, which caused the completion of the current series of recorded data, will be the first initial element that determines the next series.

On the other hand, as can be seen in Fig. 3, the measured values may not only have sharp jumps, but also remain unchanged for a long period of time. (Fig. 3b, channel 1), or the differences between adjacent readings remain constant (Fig. 3a, channel 1, individual sections on the slopes of triangular-shaped signals). This leads to a conclusion about the advisability of supplementing the compression method by storing the difference values using a method known in the literature as the run-length encoding or the RLE algorithm [3]. Its essence lies in the fact that the sections included in the registered sequence of measurement results, consisting of constant values, are replaced by a pair of numbers: one of which specifies the number of such values, and the other is equal to this value itself.

Since, while observing even one and the same process, the nature of its change in time can be completely different, it was proposed to modify the standard algorithm by introducing additional marks into the sequence of pairs "value - number of repetitions". These marks specify the number of bits allocated to record the maximum possible number of repetitions, to record the difference value, and to indicate the total number of bytes allocated to write the series of pairs before the next mark. For example, if the difference values lie within eight quantization levels, taking into account the sign (from -3 to $+4$), and the number of repetitions of identical values does not exceed 16, then $3 + 4 = 7$ bits is required to store one pair "value - number of repetitions". Then 3 such pairs can be written in 3 bytes and 3 bits will remain for recording the sign code that specifies the nature of the data layout in the following bytes. In this case, regardless of the method of arranging information into bytes, the sign bit of the highest byte must act as a flag indicating that the bits following it (from one to three) are a mark and will set the nature of the arrangement and the number of subsequent information bytes.

Continuous analysis of incoming data allowed the proposed hybrid approach to adapt the packaging format of observational results to their changing dynamics. As a result, the compression ratio became either comparable or, depending on the nature of the recording process, exceeded the compression ratio of modern archivers. The conducted studies, whose results are presented in Table 2, corroborate that without loss of information, the compression ratio of already normalized data can reach 14.2 times under favorable observation conditions and 7.7 times under unfavorable conditions [16]. At the same time, unlike classical algorithms, the method provides in-line rather than block compression, is algorithmically simple and does not require significant computing resources for implementation.

Table 2. The results of data compression of the temperature channels of the monitoring system by conventional archivers and the proposed algorithm (Org). Initial data – 14000 bytes of normalized 7000 readings. Bold font highlights the most efficient compression algorithms.

Bzip2	LZMa	LZMa2	PPMd	GZIP	XZ	RAR	ZIP	Deflate	Org2
Sudden changes in outdoor temperature									
1406	1597	1602	1723	1725	1542	1403	2137	1806	**1820**
Smooth changes in room temperature									
992	1138	1153	1220	1240	1091	1052	1561	1294	**980**

5   Lossy Compression Methods

All real measurements have errors, there is not essential to store data with a high precision if the measurement error exceeds the rounding error when recording the result. Also, it makes no sense to save the measurement result with high accuracy when, due to the presence of external noise, interference and various other destabilizing factors, it contains errors and deviations that exceed the instrumental error. This is especially evident when presenting the measurement results in digital form. For instance, in Fig. 3b, channel 1 and in Fig. 3c, channel 4, for a sufficiently long time, the measured value remains generally unchanged or occasionally changes from one quantization level to another, but at the same time fluctuates continuously within one quantization level. As a result, the use of the previously considered compression methods based on RLE algorithms becomes ineffective, since the length of sections with constant values becomes very short due to such fluctuations.

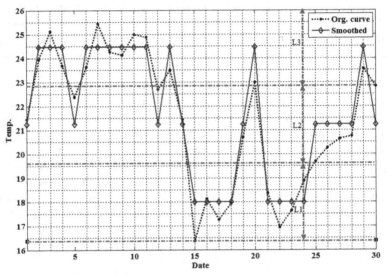

Fig. 5. Application of non-linear filtering to reduce the number of changes in the registered physical quantity.

Averaging the values of the time series over several points, for example, using the moving average method, or by approximating segments of a fixed length with polynomials or trigonometric functions, or using nonlinear filtering reduces such fluctuations. The result of applying one of the nonlinear filtering implementations described in [17] is shown in Fig. 5.

In this filtering method, the entire variation range of a physical parameter is segmented into layers with a thickness denoted as Li. Within each layer, actual measurements are approximated by a median value of that layer. To enhance the smoothing effect, a signal transition from one layer to another is performed only after a predefined number of successive signal values are registered within that layer.

Introducing a "zone of uncertainty" between layers can further improve this smoothing. When a signal is detected in this zone, it is considered that it retains the value of the layer in which it was previously located.

Figure 6 exemplifies this method. It was assumed that in order to change the reading to another layer, it was necessary to repeat three values belonging to that layer. In order to move to another layer, three values belonging to this layer must be repeated.

Experimental studies show that the use of the described compression methods can reduce the amount of stored data by more than 60 times [18].

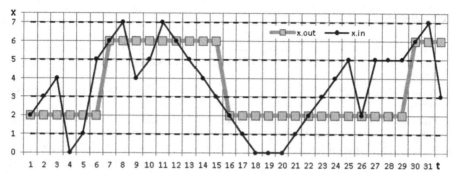

Fig. 6. Increasing the efficiency of nonlinear filtering by introducing areas of uncertainty (x = 3...5) and accumulating triple confirmations about the need to move to another area (t = 6...8; 15...17 и 29...31).

6 Conclusions

In this study, the implementation of the proposed new and modified measurement data compression methods demonstrated superior performance in data storage, compression speed, and data retrieval compared to the traditional methods outlined in [19–23]. This improvement occurs due to the fact that before the stage of applying the classical RLE algorithm and the difference value recording algorithm, the data is pre-analyzed in order to select a data storage format, and due to the packaging of data over several samples into an integer number of bytes. Additionally, markers are embedded in the data stream to identify storage formats.

The greatest compression effect is achieved when using lossy compression methods. By introducing a preliminary nonlinear filtering phase before lossy compression, fluctuations in the low-order bits of the digital representation of observations are minimized. This dramatically increases the compression ratio, from 7–10 to 50–60 times, and at the same time allows to preserve almost all the pragmatic value of the stored information.

Another promising direction in the development of monitoring data compression methods is the use of methods based on the E-layer model [24, 25]. In particular, they are most effective for real-time data compression when transmitting measurement information, as proposed in [26].

Acknowledgments. The author expresses gratitude to students, undergraduates and graduate students of Altai State Technical University named after. I.I. Polzunov, who took an active part in the creation of the monitoring system and in the implementation and experimental verification of the solutions proposed by the author and described in this work. First of all, these are Alexey Matyas, whose dissertation formed the basis of the subsequently created information-measuring system, Roman Kunz, who implemented a system for collecting and transmitting data from sensors to the SQL server, and graduate student Hussein Shawki (Minia, Egypt), who conducted most of the experimental researches and created the software necessary for their implementation.

References

1. Mukherjee, A., Kumar Panja, A., Dey, N.A.: Beginner's Guide to Data Agglomeration and Intelligent Sensing. Academic Press, eBook (2020). https://doi.org/10.1016/C2019-0-018 07-9
2. Nelson, M., Gailly, J.-L.: The Data Compression Book, 2nd edn. M&T Books, New York (1995)
3. Solomon, D.: Data Compression: The Complete Reference. Springer, London (2007). https://doi.org/10.1007/978-1-84628-603-2
4. Hahina, A.M.: Modern archiving algorithms (in Russ.) (Sovremennye algoritmy arhivacii). Zametki Uchyonogo **2**, 189–192 (2020). https://www.elibrary.ru/item.asp?id=43861716. Accessed 10 Oct 2023
5. Lihtsinder, B.Ya., Bakai, Yu.O.: Adaptive compression of measured data in wireless sensor networks. Measur. Monit. Manag. Control **1**, 52–57 (2021). http://ics.khstu.ru/journal/articles/1063. Accessed 10 Oct 2023
6. Bogachev, I.V., Levenets, A.V., Chie, Y.U.: Geometric approach to data compression of telemetry systems (in Russ.) (Geometricheskij podhod k szhatiyu dannyh telemetricheskih sistem). Informatika i sistemy upravleniya **4**(46), 16–22 (2015). https://elibrary.ru/item.asp?id=24817310. Accessed 10 Oct 2023
7. Artyshkin, A.B., Kuksenko, M.A., Pantenkov, A.P.: Economical coding as a method of increasing the speed of information transmission in telemetry systems. Bulletin of the Russian New University. Series Complex Systems Models, Analysis and Control, vol. 1, pp. 43–54 (2020). https://vestnik-rosnou.ru/node/2630. Accessed 10 Oct 2023
8. Bakulina, M.P.: Efficient coding of run length encoding during facsimile data transmission over the network (in Russ.) (Effektivnoe kodirovanie dlin serij pri faksimil'noj peredache dannyh po seti). Polzunovsky Vestnik **2**, 257–259 (2013). https://elibrary.ru/item.asp?id=202 72046. Accessed 10 Oct 2023
9. Minkin, A.S., Nikolaeva, O.V., Russkov, A.A.: Hyperspectral data compression based upon the principal component analysis. Comput. Opt. **2**(45), 235–246 (2021). https://doi.org/10.18287/2412-6179-CO-806
10. Yarmolenko, A.S., Skobenko, O.V.: Application of the theory of wavelets in the compression and filtering of geoinformation. Notes Mining Inst. **234**, 612–623 (2018). https://doi.org/10.31897/PMI.2018.6.612
11. Suchkova, L.I., Hussein, H.M., Yakunin, M.A., Yakunin, A.G.: Scalable software and hardware solutions for environmental and technical monitoring (in Russ.) (Masshtabiruemye programmno-tekhnicheskie resheniya dlya ekologicheskogo i tekhnicheskogo monitoringa). Polzunovskij Al'manah **1**, 75–82 (2013). https://www.elibrary.ru/item.asp?id=22135591. Accessed 10 Oct 2023
12. Hussein, H.M., Kuntz, R.V., Suchkova, L.I., Yakunin, A.G.: Design and implementation of weather and technology process monitoring systems. Izvestiya Altajskogo Gosudarstvennogo Universiteta **1**(77), 210–214 (2013). https://www.elibrary.ru/item.asp?id=20163999
13. Yakunin, A.G., Hussein, H.M.: Hardware-software and algorithmic provision of multipoint systems for long-term monitoring of dynamic processes. IOP Conf. Ser. J. Phys. **881**, 012028 (2017). https://doi.org/10.1088/1742-6596/881/1/012028
14. Hussein, H.M., Suchkova, L.I., Yakunin, A.G.: Ways for improving methods of data storing in monitoring systems. Polzunovskij Al'manah (Polzunovskiy Almanac) **2**, 48–50 (2012)
15. Hussein, H.M., Yakunin, A.G.: Data differencing method to optimize data storing in weather monitoring system. Polzunovskij vestnik (Polzunovsky Herald) **2**, 65–68 (2013). https://journal.altstu.ru/media/f/old2/pv2013_02/pdf/065hussein.pdf. https://www.elibrary.ru/item.asp?id=20271998

16. Melnikov, E.M., Yakunin, A.G.: Investigation of the effectiveness of lossless compression methods for temperature monitoring data (in Russ.) (Issledovanie effektivnosti metodov szhatiya bez poter' dannyh temperaturnogo monitoringa). In: Proceedings of the XX International Scientific and Technical Conference Measurement, Control, Informatization, Polzunov Altai State Technical University, Barnaul, 23 May 2019, pp. 32–38 (2019). https://elibrary.ru/item.asp?id=41166261

17. Hussein, H.M., Yakunin, A.G.: Simple curve smoothing methods for weather monitoring system. In: Materials of the XI International Scientific and Practical Conference "Science and Technology: A Step into the Future-2015", Publishing House "Education and science", Praha, 27 February–5 March 2015, pp. 73–76 (2015)

18. Hussein, H., Yakunin, A., Suchkova, L.A.: Comparison of data compression methods for solving problems of temperature monitoring. MATEC Web Conf. **79** (2016). https://doi.org/10.1051/matecconf/20167901076

19. Zhen, Ch., Ren, B.: Design and realization of data compression in real-time data-base. In: Proceedings of 2009 International Conference on Computational Intelligence and Software Engineering, Wuhan, China, 11–13 December 2009, pp. 340–243 (2009). https://doi.org/10.1109/CISE.2009.5366670

20. Ray, G.: Data compression in databases. Master's thesis, Department of Computer Science and Automation, Indian Institute of Science, June (1995)

21. Yang, H.: On the performance of data compression algorithms based upon string matching. IEEE Trans. Inf. Theory **44**(1), 47–65 (1998)

22. Lin, M., Chang, Y.Yi.: A new architecture of a two-stage lossless data compression and decompression algorithm. IEEE Trans. VLSI **17**(9), 1297–1303 (2009)

23. Tamrakar, A., Nanda, V.: A compression algorithm for optimization of storage consumption of non-oracle database. Int. J. Adv. Res. Comput. Sci. Electron. Eng. **1**(5), 39–43 (2012)

24. Yakunin, A., Suchkova, L.: Application of the E – layer model for solving the problems of parametric estimation in measuring devices. MATEC Web Conf. **155**, 01020 (2018). https://doi.org/10.1051/matecconf/201815501020

25. Suchkova, L.I., Yakunin, A.G.: Application of E-regions of the parameters of the model function of the signal for the extraordinary situations detection. In: 2018 International Multi-Conference on Industrial Engineering and Modern Technologies, Vladivostok, pp. 1–6. IEEE Xplore (2018). https://doi.org/10.1109/FarEastCon.2018.8602766

26. Suslova, S.V., Yakunin, A.G.: A method for extracting a trend of a non-stationary process with adaptation of approximation intervals (Sposob vydeleniya trenda nestacionarnogo processa s adaptaciej intervalov approksimacii). Russia Patent 2,645,273, 19 February 2018 (2018)

Object Recognition Based on Three-Dimensional Computer Graphics

Nikolay Lashchik[✉]

MIREA – Russian Technological University, Vernadsky Avenue 78, 119454 Moscow, Russia
lkolya97@gmail.com

Abstract. This paper considers the issue of training convolutional neural networks based on the representation of a data set in the form of three-dimensional computer graphics. The principle of operation of convolutional neural networks is to train and recognize categories of objects using orthogonal projections of a three-dimensional object. In the recognition process, difficulties arise in mirroring the object, which are solved by using the Pearson correlation coefficient. To use a convolutional neural network, a comparative analysis of the architectures of existing effective image recognition solutions for recording visual information was carried out. For the correct operation of the trained neural network, three-dimensional graphics objects were modeled. After obtaining the desired three-dimensional model, a black-and-white image was obtained, which was used to obtain the contours of the white spots. It is on such black-and-white drawings that the object recognition program is trained. The success of the neural network has been proven experimentally. Thus, it is possible to recognize real objects based on convolutional neural networks trained in virtual space.

Keywords: Three-Dimensional Object · Object Category Recognition · Object Mirroring · Convolutional Neural Networks · Machine Learning

1 Introduction

Object recognition is one of the most important cognitive functions of the human brain. Firstly, the recognition process begins with the process of perceiving an object. This process of perception is carried out with the help of the organs of perception, chiefly vision. Contemplation as a way of cognitive activity helps a person read information about the distinctive features of the observed objects [1].

Regarding artificial intelligence technologies, the problem of object recognition has been successfully solved by computer vision based on convolutional neural networks. With their help, two-dimensional graphical information can be manipulated and transformed. However, there are situations when it is impossible to provide a training sample in a timely manner. For example, a person has the ability to easily adapt to different situations and environments by observing new categories of objects. On the contrary, to transfer an automatic device to a new environment, it is often necessary to completely change the new knowledge base. It is obvious that many important issues in the field of

© The Author(s), under exclusive license to Springer Nature Switzerland AG 2024
V. Jordan et al. (Eds.): HPCST 2023, CCIS 1986, pp. 132–142, 2024.
https://doi.org/10.1007/978-3-031-51057-1_10

computer vision have been successfully solved, but there are tasks that require improvement in their solutions. This paper discusses the issues of training a convolutional neural network based on three-dimensional computer graphics.

2 Problem Statement

Object recognition using very limited training data is crucial for many computer vision applications. This task becomes even more difficult when the system needs to learn about new categories of objects from very few examples. It seems very difficult to use convolutional neural networks for such tasks, since they require a lot of data and are prone to catastrophic forgetfulness. However, this problem can be solved on the basis of the instance-based learning and recognition approach, which considers the study of categories as a process of studying instances of a category, i.e., a category is represented only by a set of known instances, $C \leftarrow \{O\_1,\ldots,O\_n\}$, where is a representation of the type of object based on combining layers [2].

Instance-based learning, also known as memory-based learning, is a basic approach to evaluating the representation of objects. The advantage of the instance-based approach compared to other machine learning methods is the ability to quickly adapt the object category model to a previously invisible instance by saving a new instance or deleting an old one. This approach to learning is able to recognize objects using a few exponential instances while maintaining too many redundant instances, which leads to large memory consumption and slows down the recognition speed [3]. Therefore, every time a new request appears, its previously saved data is checked, and a value for the new instance is assigned to the target object.

3 Related Work

At the moment, the following methods are used to solve the problem of insufficient training data: the first and most common method, which can help with the lack of training data, is completely avoiding test samples, but this option is the least effective method of solving this problem. Its main advantages are ease of execution and time savings. However, the consequences of such a decision are difficult to predict. If the models combining input and output variables are simple enough, the resulting neural network model may be suitable enough, but verifiability will only appear at the stage of practical work, when the cost of the solution resulting from the interpretation of the results will increase disproportionately. For complex or poorly understood models, it is better to immediately abandon this approach.

The second method is to apply the theory and methods of fuzzy logic and fuzzy sets. The idea is to create "if-then" rules to determine the relationships between different input parameters, which allows you to solve the problem of inconsistency of data and their insufficient quantity. Using these rules and methods, new examples are created that help increase the amount of available data for training. As a minus of this approach, it can be noted that the use of fuzzy logic methods can lead to a decrease in the accuracy and reliability of the model since they are based on fuzzy "if-then" rules that may be less accurate and objective than traditional machine learning methods. In addition, the creation and processing of such rules can be a complex and time-consuming process that requires deep knowledge in the fields of fuzzy logic and mathematical modeling.

The third approach offers a solution to the problem of a lack of training data by dividing input and output variables into smaller groups. Then simpler artificial neural networks - single-layer perceptrons—are applied to each of these groups. These perceptrons can be trained using the available data volumes. After learning, the perceptrons are combined into a single structure called the perceptron complex. The disadvantage of this approach may be a decrease in the accuracy of the model, since the separation of variables into groups can lead to a loss of information and a decrease in the detail of the model. In addition, the process of dividing variables into groups and choosing the optimal number of groups can be quite complicated and require a lot of experiments and testing.

4 Proposed Solution

Recently, considerable attention from the perspective of computer vision has been paid to the in-depth study approach, which can be divided into three methods depending on their input data: data based on volume, where the object is represented as a three-dimensional voxel grid, which is defined as input data for a convolutional neural network; data based on form, where the object is represented by a set of two-dimensional images by projecting object points onto a plane; and data based on a set of points where the object is represented behind a previously marked three-dimensional image. Among these methods, the most effective in the recognition of objects were the methods based on form; they allowed the researchers to achieve the best recognition results to date. At the same time, this representation greatly simplifies the study of the object in the environment of creating three-dimensional computer graphics "Blender".

Thus, the point cloud of a virtual object is represented as a set of points p_i, where $i \in \{1, \ldots n\}$. . Then each point can be represented by x, y, and z coordinates. Based on the analysis of eigenvectors, three main axes of the object are constructed, and the center of the object is determined using the capabilities of the Blender software. On this basis, orthogonal projections can be implemented to create object views. Up to six types of projections can be implemented, but three are enough to represent a three-dimensional object, including a front, top and right view (Fig. 1).

It should be noted that the direction of the eigenvectors is not unique, orthogonal projections can be mirrored. The Pearson correlation coefficient is used to eliminate ambiguity. Then the projected point $\rho = (\alpha, \beta) \in \mathbb{R}^2$, where α is the distance perpendicular to the horizontal axis, and β is the distance perpendicular to the vertical axis. The Pearson correlation coefficient r is calculated for the direction of the X axis by the formula:

$$r_x = \frac{\sum(\alpha_i \beta_i - \overline{\alpha}\overline{\beta})}{\sqrt{\sum(\alpha_i - \overline{\alpha})^2}\sqrt{\sum(\beta_i - \overline{\beta})^2}}, \tag{1}$$

where $\overline{\alpha}$ и $\overline{\beta}$ - average value α and β. According to the formula, the properties of the correlation coefficient r are determined: the Pearson correlation r varies in the range

Fig. 1. Example of an object and its orthogonal projections.

from -1 to $+1$; the value of the Pearson correlation r indicates how close the points are to a straight line.

In particular, if Pearson correlation $r = +1$ there is an absolute positive correlation, and if Pearson correlation $r = -1$ there is an absolute negative correlation. Similarly, the indicator is calculated r_y for the Y axis using a plane YoZ. Further, the sign of the axes is defined as $s = r_x r_y$, where s can be either positive or negative. In the case of a negative value of s, the three projections should be mirrored, while in the case of a positive value of s, they should not.

When the dissimilarity of the new object with all previously known categories exceeds the threshold value, the automatic device concludes that this object does not belong to known categories and thus initializes a new category marked as "category_m $+ 1$", where m is the number of currently known categories. Moreover, in the case of a similarity measure, the difference between two objects can be calculated using various distance functions. For example, Pearson chi-squared is used to estimate the similarity of two instances.

When processing each image, it is necessary to implement the detection of the contours of objects (Fig. 2).

As a result, an image will be obtained (Fig. 2) on which all the objects you are looking for will be white spots and the background will be completely black. Next, you can use the built-in OpenCV library's findContours function to obtain the contours of white spots representing the desired objects.

A convolutional neural network can be considered a special kind of artificial neural network adapted for effective pattern recognition in an image. It implements some features of the visual cortex of the brain, in which two types of cells were discovered: simple, reacting to straight lines at different angles, and complex, whose reaction is associated with the activation of a certain set of simple cells. There are many implementations of neural networks of this type for classifying objects in images and image recognition.

Among the tested neural networks, the models of the YOLOv3 group look the most attractive, namely YOLOv3–320, YOLOv3–416 because of their high performance (> 30 FPS) and accuracy that is not lower than the average of the others (> 50%). The model is based on the Darknet framework. Despite the different names, in fact it is the

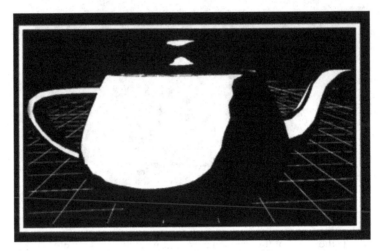

Fig. 2. The result of image processing.

same model, and the addition of 320 and 416 in the name denotes the width of the image coming to the input.

A 448 × 448 image is fed to the network input, which, in case of inconsistency, will be stretched or compressed to this size. The network consists of 106 layers and uses only 4 types of layers and 2 types of operations on layers (Fig. 3).

The first type is a convolutional layer. It creates a feature map based on the input data, in which the main patterns found in the image are highlighted. The second layer is the routing layer, which is necessary to determine the two layers, the result of which must be taken in order to perform the merge operation. The third type is the superselection layer. It increases the output dimension due to the feature map obtained from the convolutional layer. This is necessary for the correct execution of the merge operation. The fourth type is the detection layer. It is used to determine the location of the framing windows. The union operation concatenates feature maps from two layers, and the addition operation sums feature maps element by element.

Before training, it is necessary to configure the network configuration. To do this, it is necessary to set the number of classes equal to the original on the detection layers and, on the convolutional layers preceding them, change the dimension of the feature map to a value equal to the number of classes [4]. At the same time, at each epoch, the original images were subjected to various modifications, for example, turning the image by a small angle, changing the color scale, etc. In order to achieve better recognition quality, the weights of an already trained neural network were used on a set obtained in the software for creating three-dimensional computer graphics "Blender". YOLOv3 uses the sum of root-mean-square errors as an accuracy metric

$$E_{MSE} = \frac{1}{n} \sum\nolimits_{j=1}^{n} \left(d_j - y_j\right)^2. \tag{2}$$

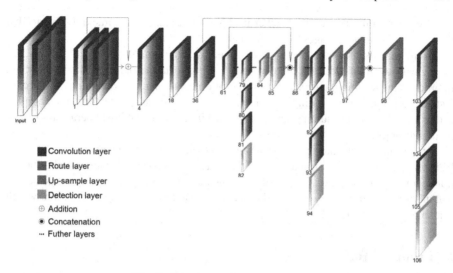

Fig. 3. YOLOv3 network architecture.

In machine learning theory, learning error is understood as the difference between the desired (target) d_j and the actual y_j output of the model using examples of the training set with the number of observations n.

When developing an automatic system based on computer vision, it is not enough to implement only the search for objects in the frame. Since a continuous stream of real-time video is received at the input, it is also necessary to learn how to programmatically track the movement of found objects. To do this, object trackers come to the rescue. An object tracker is an algorithm that determines the location of a moving object (or several objects) over time. Tracking a moving object can be a time-consuming process due to the amount of data contained in the recording of visual information (video). The complexity is further compounded by the possible need to use object recognition techniques for tracking, which is itself a difficult task [5].

The most popular when the movement is predictable and small is the MedianFlow tracker. The principle of operation of this tracker is as follows [6]:

1. The algorithm splits the image of the tracked object into small fragments
2. For each fragment, a new position is found on the next frame by calculating the optical flow between the two images.
3. To estimate the new position of the object, the tracker calculates the median of the displacement vectors of all image fragments.
4. To estimate the change in the size of the tracked object, the tracker calculates the median of the distances between all fragments of the image.

The tracker tracks the object both forward and backward in time and measures the discrepancies between these two trajectories. Next, the tracker selects the trajectory with the minimum error (difference) between the two frames. Unlike other trackers that continue to work and try to track an object even when it is either "lost" by the tracker or disappears from the frame, this tracker is able to determine when tracking failed. This is one of the main advantages of this tracker.

An increase in the objects being determined by an automatic system can allow identifying various situations at the level of visual-effective and visual-imaginative thinking. In order to move from the usual recognition of objects based on perception, in our case visual, to recognition by finding various patterns, it is necessary to perform the complex and extremely time-consuming work of identifying various patterns characteristic of a particular object.

At the same time, data sets for training a convolutional neural network can be taken from virtual reality, and testing is carried out in a real environment using visual information recording. This approach, according to the authors, will create an automatic device that allows not only to recognize real objects but also to determine the situation, which is also caused by virtual exposure.

5 Proof of Work

After the implementation of the architecture and algorithm of the convolutional neural network, it becomes necessary to verify the correctness of the implemented program.

First and foremost, you need to make sure that the lack of training data is really a critical problem when training neural networks. To prove this statement, the neural network was trained to recognize a teapot on a small amount of data. In this particular experiment, 10 photos of a teapot were taken, although it is believed that the minimum number should be at least 100.

After training the neural network, photos of dummies from a different set, different from the one on which the network was trained, were transmitted. Out of a hundred photos, a neural network trained on an insufficient amount of test data correctly identified the subject only 20 times, which, as it is easy to calculate, is only 20% of successful test results.

After receiving the result with which the work of the program proposed in this article will be compared, it is necessary to implement a simple and intuitive interface that would show the work of the program. It is necessary to conduct tests not only to prove the correct operation of the neural network, but also for cases where incorrect data for analysis and objects unknown to the program were provided. For these purposes, a program was implemented to check the definition of the object depicted in the photo. This program has a simple window interface (Fig. 4), in which you need to enter the path to the file with the image of the object to be recognized. After clicking on the confirmation button, the program should output a message with the name of the object that the neural network recognized in the photo.

If you enter the wrong path to the file, a message is displayed stating that the wrong path to the image is specified. This reaction occurs when you enter a path where no file exists or if a file with an extension other than the extension that is an image file is located at the specified address.

If an image is submitted to the input of the program that does not depict a teapot, then, as expected, the program cannot recognize this object and issues a message about it.

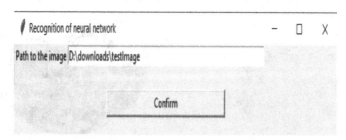

Fig. 4. Interface of the window application for checking the operation of the program.

After all possible variants of the unsuccessful outcome of the program have been checked, you can test the reaction of the neural network to the picture showing the teapot.

For the recognition test, a photo of a real teapot was taken (Fig. 5), which looks like a three-dimensional model of a teapot on which the neural network was trained.

Fig. 5. The photo on which the first test of the program was carried out.

After launching the program, it is clear that the neural network correctly recognized that the teapot is located on the entered image.

After the initial successful test, another image of the teapot was selected (Fig. 6a), which is slightly different from the first test image and the model on which the program was trained. The photo taken differs from the first one in the presence of gold edging on the upper parts of the teapot and its lid.

After launching the program, a response was received that the neural network recognized a teapot in the input image. This test shows that the presence of some additional colors on the tested image does not affect the correctness of the program output. For the third test, a photo of a teapot of a different color with patterns applied to it was selected (Fig. 6b).

The program responded to the entered photo by recognizing a teapot in it. The test can be considered successful. He shows that a trained neural network can recognize teapots not only in white but also with slight differences from the images that were

a)   b)

Fig. 6. Sample objects in the images used for the second and third tests: a) teapot that is quite similar in design to those used in training set; b) teapot that slightly varies from those used in training set (features complex color patterns).

used in the training of artificial intelligence. The developed program is able to recognize objects of different colors with various additional images on them.

The fourth test was conducted on the basis of a photo of a teapot, which is very different from the object of the training (Fig. 7a). In addition to the presence of a handle, it is also distinguished by its shape and a shortened spout. Despite the noticeable differences in the test object, the program was able to recognize a teapot in it.

A mug photo was selected for the final demonstration test (Fig. 7b). This choice was made due to the fact that the mug is close to the teapot in meaning but nevertheless differs in its functionality and representation.

a)   b)

Fig. 7. Samples of objects in the images used for the fourth and fifth tests: a) teapot that is significantly different from the training set; b) fancy cup with partial visual resemblance of a teapot.

On the entered photo, the program gave an error that it could not recognize the object that is depicted. This test shows that a trained neural network correctly recognizes only the objects on which it has been trained. Thus, it is possible to perceive this test as passing positively.

Table 1. Results of quantitative evaluation.

Object type	Number of tests	Accuracy
Teapot similar to training set (as in Fig. 6a)	100	93
Teapot that varies slightly from training set (as in Fig. 6b)	100	87
Teapot different from the training set (as in Fig. 7b)	100	79
Different object of houseware (as in Fig. 7a)	100	99

Next, a larger-scale testing of the neural network was carried out, in which images of teapots were transmitted to the program for recognition, as similar as possible to the objects on which it was trained, slightly different from them, and very different (Table 1). 100 photos were selected for each category of test objects to be able to approximate the percentage of correct operation of the program. Each group was selected according to the same criteria, namely, image quality, that is, photos were selected in which objects occupying most of the picture are clearly visible. It should be clarified that the dataset does not pretend to be the result of extensive and detailed testing but is simply used to approximate the results of the program since there is no ready-made solution for such testing.

Testing was also carried out, in which other kitchen implements were transferred for which the program was not trained. In this case, the successfully passed test is the one in which the neural network did not recognize the teapot.

After the tests of the program, it can be seen that the developed and trained neural network correctly recognizes the teapot in the photos, which means that this approach to training an artificial network on three-dimensional image models is effective.

Naturally, if a teapot is depicted in the photo but is of a different shape, then the neural network will be worse at determining the ownership of this object. Moreover, the more differences there are, the greater the chance of an incorrect definition of the object. This problem can be solved by increasing the variety of object shapes in a variety of training data.

6   Conclusions

Since the developed and trained convolutional neural network has been experimentally proven to successfully recognize 3D models, its further development makes it possible to obtain the result of processing not just one 3D model but also several 3D models located on the same image. By increasing the variety of possible forms and characteristics of training data, it is possible to improve the correct recognition of objects, even if they are very different from the usual representation of the training object.

Then, on the basis of the objects being determined, it is possible to determine various situations using imaginative thinking at the program level. As already mentioned, in order to move from the usual recognition of objects based on perception, in our case, visual, to recognition by finding various patterns, it is necessary to perform complex and extremely long work just to identify various patterns characteristic of a particular object. Neural networks are great for this kind of task. With properly structured neural network training, the time for this very training is enormously reduced if we compare the time for which a person and a computer will be able to determine a set of attributes related to a specific object.

In the future, it is planned to develop such an artificial intelligence system that will be able not only to recognize all the objects in the image but also to determine the situation that occurs at the exposition. For example, train the system to recognize not only the teapot but also the cups located next to it at the same time. After correct recognition, the program can make the assumption that in this situation, the contents of the teapot can be poured into mugs.

References

1. Shimonishi H., Murata M., Hasegawa G., Techasarntikul N.: Energy optimization of distributed video processing system using genetic algorithm with bayesian attractor model. In: Proceedings of 2023 IEEE 9th International Conference on Network Softwarization (NetSoft) (2023). https://doi.org/10.1109/NetSoft57336.2023.10175483
2. Cha, S.-H.: Comprehensive survey on distance/similarity measures between probability density functions. Int. J. Math. Models Methods Appl. Sci. 1(4), 300–307 (2007)
3. Hamidreza Kasaei, S.: OrthographicNet: a deep transfer learning approach for 3D object recognition in open-ended domains. IEEE/ASME Trans. Mechatron. 26(6), 2910–2921 (2021). https://doi.org/10.1109/TMECH.2020.3048433
4. Golovko, V., Kroshchanka, A., Mikhno, E.: Brands and caps labeling recognition in images using deep learning. In: Ablameyko, S.V., Krasnoproshin, V.V., Lukashevich, M.M. (eds.) PRIP 2019. CCIS, vol. 1055, pp. 35–51. Springer, Cham (2019). https://doi.org/10.1007/978-3-030-35430-5_4
5. Anto Bennet, M., Srinath, R., Abirami, D., Thilagavathi, S., Soundarya, S., Yuvarani, R.: Performance evaluation of video surveillance using mete, melt and nidc technique. Int. J. Smart Sens. Intell. Syst. 10(5), 25–54 (2022). https://doi.org/10.21307/ijssis-2017-234
6. Lukezic, A., Vojr, T., Zajc, L.C., Matas, J., Kristan, M.: Discriminative correlation filter tracker with channel and spatial reliability. Int. J. Comput. Vision 126(7), 671–688 (2018)
7. Kochanov, A., Zolotukhin, V., Mironenko, V., Savelyeva, A., Polyakova, A.: Digital processing of satellite images using neural network algorithms. J. Phys. Conf. Ser. 2373, 062026 (2022)

Adaptive Methods for the Structural Optimization of Neural Networks and Their Ensemble for Data Analysis

Vladimir Bukhtoyarov$^{(\boxtimes)}$ ⬛, Vladimir Nelyub⬛, Dmitry Evsyukov⬛,
Sergei Nelyub, and Andrey Gantimurov

Bauman Moscow State Technical University, Moscow, Russia
bukhtoyarov@emtc.ru

Abstract. The article is devoted to studying the effectiveness of ensemble methods of neural network solvers for solving a regression problem. The problem of ensuring the efficiency of using a prediction model based on a set of artificial neural networks, which should provide increased efficiency compared to a single regressor, is considered. One of the requirements to ensure an increase in efficiency is that the solvers in the ensemble must be sufficiently different to ensure that the ensemble model exits the region of the local minimum error of a single solver. For the considered case of constructing neural network ensembles, it is proposed to provide distinction based on the generation of different structures of neural network regressors. To do this, a modification is introduced into the structure of the previously developed probabilistic method for designing neural networks. This modification is based on the use of special coefficients that determine the adaptation of the probability of using various activation functions depending on the presence of those in already formed neural networks in the ensemble. The proposed approach was implemented in a software system and tested using generated datasets and real industrial data sets, described in the article. The results obtained indicate the relatively high efficiency of constructing collective regressors when forming neural networks using the proposed approach while maintaining diversity in conditions of noisy samples.

Keywords: Modeling · Simulation · Artificial Neural Networks · Ensembles · Regression · Structure Shaping · Probabilistic Approach

1 Introduction

Today, the use of artificial neural networks makes it possible to effectively solve problems of data mining, such as modeling, classification, and clustering. In the course of the ongoing research, regression modeling problems were considered, involving the use of artificial neural networks to describe the dependencies between input and output data. There are quite a few applications of artificial neural networks in this focus, including in metallurgy, mechanical engineering and medicine [1–3].

© The Author(s), under exclusive license to Springer Nature Switzerland AG 2024
V. Jordan et al. (Eds.): HPCST 2023, CCIS 1986, pp. 143–157, 2024.
https://doi.org/10.1007/978-3-031-51057-1_11

However, we can observe a constant increase in the complexity of data analysis problems. The technological processes themselves and the models that describe them are becoming more complex. This is due to the fact that there is an increase in the number of controlled process parameters and, accordingly, an expansion of the factor space for searching for effective models. There is also an increase in the amount of data processed to build regression models, and the demands on the accuracy of such models are becoming more and more demanding. All this leads to the fact that the process of searching for methods that would allow the formation of more efficient regression models, including criteria for computational efficiency and accuracy of describing dependencies in data, is constantly ongoing. This is true for both regression neural network models and other types of models.

Among several directions for improving approaches to building regression models (and, in general, classifiers), one can single out ensembling. This approach involves the construction and integration of models and classifiers into groups that process data sets more efficiently than with an individual regressor or classifier [4–6]. A limitation of the approach is the fundamental complexity of the ensemble compared to an individual solver. In this regard, the developed schemes for constructing ensembles should be effective in terms of the use of individual solvers. Some of them, in particular, implement the requirements of maintaining diversity in such an ensemble of solvers.

One of the fundamental works in the field of artificial neural ensembles is [7]. The authors have shown an increase in the efficiency of the recognition system when using several neural network classifiers instead of a single artificial neural network. This approach and its variants were refined and implemented to solve practice-oriented problems, including in the field of recognition, medical and technical diagnostics, processing of seismic monitoring signals and many others [8–11].

Despite the demonstrated possibility of using ensembles for certain practical problems, this approach is associated with an increase in the complexity of the model design stage. This is due to the need to build not one, but several at once, sometimes a significant number of neural network solvers, which are required to be effective jointly, and not separately.

There are two steps required to build the ensemble solver. At the first step, it is necessary to create individual solvers that are sufficiently effective and differ from each other (artificial neural networks are considered within the framework of this study). At the second step, it is necessary to assemble an ensemble scheme for obtaining a general solution from individual solvers, namely, their individual solutions. It appears that increasing the efficiency of solving a problem depends on each of the two steps indicated. Thus, the effectiveness of the ensemble approach is determined by the quality of the solvers in the pool formed at the first step, and the effectiveness of the method for calculating the ensemble solution from individual solutions. [6, 12–14]. Different approaches for performing the above steps can be identified in real-world applications. A complete classification is not the purpose of this review, but to determine the location of the presented study, the following must be indicated. Individual classifiers can be built over the entire dataset, or they can be defined on individual subsets of the original dataset. In this study, we considered the concept of the entire dataset for all individual solvers. The construction of a general solution can be based on simple rules such as weighted summation or median, or it can be produced in the form of a more flexible, potentially nonlinear relationship. As part of this study, we integrated into the proposed approach a scheme for generating a general solution based on genetic symbolic programming.

Further, in Sect. 2, a description is given of the proposed methods for the formation of artificial neural networks for their inclusion in the ensemble and, additionally, the method of forming the general solution of the ensemble based on the predictions of individual neural networks. The proposed methods are based on a generalized principle underlying evolutionary heuristic algorithms. Section 3 presents the results of a numerical experimental study of the proposed approaches and a comparative analysis with several alternative ensemble approaches and individual models. Section 4 contains a general conclusion on the article.

2 Materials and Methods

2.1 Designing the Structure of Neural Networks Using the Probabilistic Approach

In the case of using an ensemble approach, special attention should be paid to the stage of generating individual solvers. Such solvers should not be completely untrained, but should have some basic performance that demonstrates their adaptation using the dataset. On the other hand, they should not be overtrained, since there is a possibility of bringing them to some general local optimum in terms of performance criterion for a specific problem. In this sense, they must be different, covering different points in the solver space. A pool of such preliminary predictions may be intended to be used in whole or in part with subset selection. The scheme for the selection of ensemble members will be considered in the next section. In this section, we will consider the proposed approach to the formation of structures of artificial neural networks for subsequent training.

Since the ensemble approach assumes the need to build a sufficiently large pool of individual solvers with a high degree of confidence, the conditions of real application require automation of this process. This is due to the fact that building even one solver is quite a serious task for practical application on datasets of significantly different types. And here this problem of optimal design of an efficient solver must be solved many times with some restrictions on the complexity and variety of individual solvers. This fully applies to the use of artificial neural networks as individual solvers, which are the focus of this study. The issue of adapting neural network solvers can be most effectively resolved in the case of optimal selection of the structure and further effective training of the corresponding network. It is almost impossible to rationally determine in advance the non-redundant efficient structure of a neural network designed for a huge number of real-life datasets. The requirement of non-redundancy is essential for the ensemble approach, since the use of many redundant individual solvers can lead to a significant increase in the need for computing resources compared to the increase in the quality of the resulting solution. Limits on available computing resources may also be exceeded, which is still typical for a large number of real-world systems.

In order to further describe the method for maintaining the diversity of neural network solvers, we provide a brief description of the previously developed probabilistic method for designing the structures of artificial neural networks. This approach is for automatically designing the structure of neural networks is based on the calculation and use of probability estimates (Formula 1).

$$p_{i,j}^k, i = \overline{1, N_l}, j = \overline{1, N_{neuron}}, k = \overline{0, N_F}, \tag{1}$$

where i is the number of the hidden layer of the neural network; j is the number of the neuron on the hidden layer of the network; N_l is the maximum number of hidden layers; N_{neuron} is the maximum number of neurons on the hidden layer; k is an identifier whose value is interpreted as follows:

1. If $k = 0$, then $p_{i,j}^0$ is an estimate of the probability that the j-th neuron is absent on the i-th layer of the network;
2. If $k \in [1, N_F]$, then $p_{i,j}^k$ is an estimate of the probability that the j-th neuron exists on the i-th layer of the network and its activation function is the activation function with number k from the set of activation functions available to the algorithm. N_F is the power of the set of activation functions that can be used when designing the neural network structure.

The formula for calculating the total probability of the presence or absence of a neuron at a place in the neural network, determined by the number of the layer and the number of the conditional position of the neuron on the layer is as follows:

$$p_{i,j}^0 + \sum_{k=1}^{N_F} p_{i,j}^k = 1. \tag{2}$$

For details see [15].

2.2 Providing Diversity for Individual Neural Network Solvers

One of the priorities for building effective ensemble models is to ensure the diversity of individual solvers, ensuring the possibility of their synergistic interaction and not being biased towards the local and insufficient maximum efficiency of a single model. Maintaining diversity can be achieved in a variety of ways. This can be building or training models on different subsets of the original dataset, as is implemented in boosting schemes. Another option is to build models with initially different structures. For artificial neural networks, the focus of this research, this means designing and further training neural network solvers with different activation functions and hidden layer configurations.

Within the framework of this study, just such an approach was considered with a focus on the formation of an ensemble based on artificial neural networks. The modification of the method of probabilistic design of neural network structures is as follows. An additional coefficient is introduced that is used to correct the probability of forming the structure of a neural network with each type of activation function used. Obviously, maintaining diversity among neural networks is also ensured by their structural differences. Correcting the probabilities of the procedure producing neural networks should reduce the likelihood of the appearance of neural networks with a structure similar to those that have already been formed and placed in the pool of solvers. It is proposed to use modified probability values calculated according to Formula 3:

$$\tilde{p}_{i,j}^k = d_i^k \cdot p_{i,j}^k, \tag{3}$$

where $p_{i,j}^k$ i and iterators are defined by Formula (1) The value of the coefficient is calculated by Formula 4:

$$d_{i,j}^k = 1 - \frac{n_{i,j}^k}{Ensemble\ Size},\qquad (4)$$

where n is the number of neural networks, with a neuron of the k-th type on the i-th layer in the j-th position, already placed in the ensemble. The coefficient of change in the probability of occurrence of an activation function of a certain type d_i^k is calculated individually for all neurons of the hidden layers of the designed neural network. Possible values of such coefficients are in the range (0; 1]. If there is no neural network with the corresponding activation function of the neuron in the position of the hidden layer in the generated set of individual solvers, the coefficient is equal to 1. When such neural networks of networks appear, the coefficient decreases and, accordingly, the probability of generating a network with the same neuron in the neuron position in the hidden layers corresponding to the coefficient decreases. This is aimed at reducing the repeatability of the structure of neural networks, which is a factor in maintaining the diversity of individual neural network solvers for the ensemble model. The starting values of the coefficients for constructing the initial neural network are to equate all coefficients to 1.

The approach under consideration in a general formulation was proposed and partly studied, and the results of such a study are given in [16]. However, in the first version of the proposed approach, two important steps were not implemented, which were used and already tested on the new test function number 5, introduced in Sect. 3.

In order to increase the stability of the approach, we introduced and implemented the following modifications, compared to [16]. The formula for calculating the coefficient used to take into account diversity has been changed, which allows us to take into account the number of neurons of a certain type in already formed neural networks $N_{i,j}^k$, , and not in relation to the entire ensemble size (Formula (5)). This is due to the following factor. In the proposed evolutionary and some alternative ensemble approaches the choice of neural regressors directly used in the formation of the ensemble and its general solution occurs at the next step. Therefore, it is not possible to determine the size of the ensemble in advance. The approach used previously involved dividing by the ensemble size (for approaches with a fixed number of solvers in the ensemble) or dividing by the number of networks in the preliminary pool. This gave rise to ambiguity and dependence of the first step of ensemble formation on the second step. The proposed modification eliminates such ambiguity and makes the approach invariant from the choice of approach to forming a general ensemble solution.

$$d_{i,j}^k = 1 - \frac{n_{i,j}^k}{N_{i,j}^k}.\qquad (5)$$

In the modified version of the approach proposed here, an unconditional minimum probability was introduced, less than which the probability cannot be decremented using Formula (3). This is due to a motivated relaxation of the calculation rule using decrementation coefficients according to Formula (5) so that it is possible to obtain neural network structures of any configuration at each iteration of designing the preliminary pool or the ensemble. In this study, we used an unconditional minimum value of 0.05 for cases where recalculation of the probability according to Formula (3) leads to a $\tilde{p}_{i,j}^{k}$ value less than the established value of 0.05.

Taking into account the basic statistical conditions, in the new version of the approach to maintaining diversity in ensembles, to ensure the stability of the approach after recalculating the probabilities using Formula (3), a normalization stage has been introduced, bringing the sum of probabilities to a value equal to 1:

$$\tilde{p}_{i,j}^{k} = \frac{\tilde{p}_{i,j}^{k}}{\sum_{k=0}^{N_f} \tilde{p}_{i,j}^{k}}. \tag{6}$$

This step was not introduced and used for the initial version of this approach and had the potential to fail to ensure the completeness of the sum of probabilities, potentially reducing the effectiveness of the approach. Together with minor corrections in iterators, the introduced changes allow us to claim a significant refinement and improvement of the original approach, requiring confirmation using a new test dataset. The improved new approach was also tested using a new, more complex test function with varying levels of imposed noise.

2.3 Ensemble Design Using Modified Genetic Programming

The formation of an ensemble can be carried out in the following options. The first option is that the ensemble consists of all individual solvers. In this case, individual solvers are formed depending on the qualities of previous members of the ensemble or the explicitly or implicitly defined areas of influence of each solver. In particular, this can be attributed to various boosting schemes, which also cover the design stage of individual solvers [10, 17]. An alternative option is to compile an ensemble of solvers selected from a preliminary pool - either a small subset or all of the pool can be selected. An alternative option is to compile an ensemble of solvers selected from a preliminary pool - either a small subset or all of the pool can be selected. Such selection can be carried out in an explicit combinatorial form, which, however, is a computationally intensive process given the large number of combinations with a fairly large size of the designed ensemble. Such selection can also be carried out implicitly, for example, by resetting the weighting coefficients of individual solvers to zero in the formula for calculating the general solution, or even during the construction of the very principle of calculating the general solution, which may not take into account the predictions of all individual solvers. [18–21]. This leads to the fact that one of the problems of effective application of ensemble methods is the choice of a method for forming a general solution. In the basic case, there are two possible approaches. The first one assumes, for the regression problem, the use of some formula that is static. Such methods may include methods for calculating the average value based on the solutions of individual solvers. In addition to simple averaging, a similar scheme with the introduction of weighting coefficients can be used. Such weighting factors may be proportionally related, for example, to an estimate of the error of the corresponding individual solver. However, the issue of choosing coefficients remains, since an individually efficient solver may have too much weight to bias the ensemble solution relative to it, thereby reducing the ensemble effect. An alternative option is to use adaptive schemes for integrating and post-processing the opinions of individual classifiers. In the limit, this can be another level of solvers that accept as input the outputs of individual solvers of the first level. Keeping in mind the possibility of such an approach, the study attempted not to permanently seal the black box of an ensemble of neural networks in this form.

Therefore, we have proposed and tested a method that allows us to adaptively form a single second-level solver in the form of a formula that explicitly symbolically connects the outputs of individual solvers and the general ensemble solution. The scheme for using ensembles of neural networks suggests that the calculation of the general ensemble solution is based on the solutions obtained by individual neural networks, that is, the general ensemble solution is a certain function that depends on the predictions of individual neural networks (5):

$$o = f(o_1, o_2, \ldots, o_n), \tag{3}$$

where o is the general solution, o_i is the individual solution of the i-th network, n is the number of networks in the ensemble. To solve this method of computing an ensemble solution, we propose to use an approach based on genetic programming [22]. To solve this method of computing an ensemble solution, we propose to use an approach based on evolutionary heuristics called genetic programming. It is successfully used to construct optimal approximations, and the explicit formulaic form of the solution will allow one to analyze the composition and evaluate the influence of individual solvers when calculating the overall solution.

The standard genetic programming method is used to design symbolic formulas involving independent variables and constants. Since in the considered ensemble scheme the input values of the generated rule for calculating the general ensemble solution are the output of individual neural network models, a modification of the standard approach is required. For this purpose, a modification of the genetic programming method with a modified terminal set $T = \{o_1, o_2, \ldots, o_n, C\}$ is used. Here o_i is an individual solution of the i-th solver, n is the number of networks in the ensemble, C is a set of constants (numerical coefficients). A hybridized version of the genetic programming method was used in our study to fine-tune the efficiency-critical second stage of ensemble design. Such hybridization consists in the use of algorithms for adjusting the coefficients of a formula that describes a combination of solutions of individual solvers to determine the overall ensemble solution. An important positive aspect of using such evolutionary heuristics as a method of genetic programming is the possibility of flexible formation of a quality function for the process of evolutionary improvement of solutions. So, in the form of a convolution or multi-criteria formulation, such an ensemble efficiency criterion can include not only the values of the model or classifier deviation estimates, but estimates of the number of individual solvers used or their total complexity of neural structures or other types of solvers. This can be critical in cases of limited computing resources and the desire to maintain the speed of inference of the ensemble model compared to individual solvers.

3 Numerical Experiments

To evaluate the effectiveness of the proposed approach, numerical studies were performed on the generated sets of test problems and data sets of real problems hosted in the Machine Learning Repository [23]. The description of the generated and test sets is given below. As a unique objective of this study, a set of data obtained from the metallurgical production at the stage of obtaining end products in plums was used to evaluate the quality of regression modeling. The name of the company that provided the data and the characteristics of the dataset are not disclosed in order to ensure non-disclosure of commercial information.

3.1 Simulated Datasets

For the initial evaluation of the work of the considered methods, the generated data sets were used. The formation of datasets was carried out by randomly generating inputs and calculating the values of functions in the domain of definition and according to the formulas presented in Table 1. In previous studies, the use of such data sets made it possible to assess the adequacy of their use for assessing the effectiveness of regressors with basic indicators of the effectiveness of methods for constructing regression solvers on a number of real data sets. It seems reasonable to us to use such synthetic modeling problems for the purpose of controlled imposition of noise on data to assess the robustness of modeling methods.

3.2 Test Problems

To test the methods considered and proposed in the study for solving regression public datasets hosted in the Machine Learning Repository were also used. We used the Concrete Slump Data Set because it is a widely used publicly available data set that will allow external verification of the results obtained in this study. This data set consists of patterns describing the tensile strength tested for special samples. Such samples are made using concretes with controlled levels of inclusions of various ingredients. The total number of examples in the dataset is 103 observations. A dataset option with an expanded set of observations up to 1030 is also available. In this study, a basic 103 instance dataset was used, directly obtained from field samples.

A data set was also used that describes the implementation of the ore-thermal smelting process with various input parameters characterizing this metallurgical process. The regression modeling problem statement is as follows. Based on observations of a real object, data samples were generated that characterize the efficiency of an ore-thermal smelting furnace. Electrical parameters and the loading of other components are used as control parameters (input influences), since it is these input parameters that influence the processes within the processes, in addition, they provide the ability to continuously obtain and reliable information about them.

As input parameters of the process occurring in the furnace, the following process indicators were recorded: the amount of agglomerate loaded into the furnace; the amount of silica loaded into the furnace; the amount of coke loaded into the furnace; the amount of converter slag loaded into the furnace; electricity input; deepening of electrodes;

Table 1. Functions for generating test data.

Test problem	Simulated Function	Range	Sample size
1	3-d Mexican-Hat: $y = \dfrac{\sin\sqrt{x_1^2 + x_2^2}}{\sqrt{x_1^2 + x_2^2}}$	$x_i \in [-2\pi, 2\pi]$	1000
2	Friedman 1: $y = 10\sin(\pi x_1 x_2) + 20(x_3 - 0.5)^2 + 10x_4 + 5x_5$	$x_i \in [0, 1]$	1000
3	Friedman 2: $y = \sqrt{x_1^2 + (x_2 x_3 - \frac{1}{x_2 x_4})^2}$	$x_1 \in [0, 100]$ $x_2 \in [40\pi, 560\pi]$ $x_3 \in [0, 1]$ $x_4 \in [1, 11]$	1000
4	Gabor: $y = \frac{\pi}{2} exp\left[-2\left(x_1^2 + x_2^2\right)\right] cos[2\pi(x_1 + x_2)]$	$x_i \in [0, 1]$	1000
5	Multi: $y = 0.79 + 1.27x_1 x_2 + 1.56x_1 x_3 + 3.42x_2 x_5 + 2.06x_3 x_4 x_5$	$x_i \in [0, 1]$	1000

voltage; current strength; specific energy consumption. These parameters at the upper level make it possible to evaluate the energy and technological characteristics of the operating process of such installations and, in general, to judge the efficiency of the furnaces. In order to build a regression model, the nickel content in the waste slag in percent was selected as an output parameter.

3.3 Pool of Methods

The proposed approach was evaluated in terms of the efficiency of building a regression model by comparison with a number of alternative methods. Such alternative methods were implemented in software and used to build models in the considered test problems. The following methods for constructing individual regressors were investigated: artificial neural networks, regression support vector machine, multidimensional adaptive spline method [24, 25].

GASEN method based on the use of a genetic algorithm to select networks from a pool, and a boosting gradient method were considered as alternative ensemble methods [26–28]. Verification of the correctness of the implementation and pre-setting of parameters for the approaches under consideration were performed using this software system during a preliminary study on a basic set of test problems (Table 1).

3.4 Raw Results Processing

To form a statistically stable estimate of the results, a 5-fold cross-validation scheme with 5-fold resampling of the sample (the so-called 5-by-5 scheme) was used. For each method, sets of 25 regression model accuracy estimates were obtained. The obtained

R^2 values were averaged over all runs. We also considered the spread in the quality of the solutions obtained using a statistical estimate of the dispersion of the obtained R^2 values.

The inclusion of various methods for constructing regressors in the study led to the need for top-level equalization of the resources allocated to the method for processing the initial dataset in order to build a regression model. For this purpose, during the preliminary study, measurements of processor time were used, which is necessary for the average computational cycle of setting up the model. The parameters of the approaches used were selected to equalize this indicator. At the same time, the available parameters of the methods were adjusted on each of the tasks to ensure their optimal (taking into account processor time limitations) efficiency on each of the tasks.

3.5 Results and Discussion

ANOVA methods were used to assess the statistical significance of the results. We used the R^2 indicator as a basic indicator for evaluating methods and comparing the effectiveness of the described regression tasks. The research results are presented in Tables 2–4. Analysis of the results obtained leads us to the conclusion that non-ensemble methods are quite effective on relatively simple test problems without noise, since their approximation ability is sufficient. According to our observations, the problems on which such results were obtained include problems numbered 2, 3 and problem Concrete Slump problem. On them, the results obtained by individual solvers and ensemble solvers using several neural network models turned out to be statistically indistinguishable. The results of individual and ensemble solvers on problem 3 are somewhat different, but such a difference is very small and is not a reason for complicating the model in the form of a ensemble.

Among the considered set of test problems, the most complex are the problems of ore-thermal smelting and problems numbered 1 and 4. On such problems, the best results were obtained when using approaches based on ensembles of neural network solvers. Among these ensemble approaches, statistically significantly higher effectiveness corresponds to those that used an ensemble diversity approach. Our assumption is that methods based on single solvers may have reached their efficiency limit in the used configuration. It is the ensemble approach, built even on simpler individual solvers, that makes it possible to overcome this limit when constructing regressors in this case.

Assessing the robustness of the approaches was the subject of further experiments using noisy samples. The maximum noise level was the level indicated below from the value of the test functions used at each point of the definition domain. Table 3 presents the results of the study with a noise level of 10%. As follows from the results, the drop in the quality of approximation for non-ensemble solvers in some cases turned out to be even greater than the level of sample noise. The results of the ensemble solvers turned out to be more suitable for samples with noise simulating those in the measurement channels.

The use of ensemble methods made it possible to achieve more stable results, although a decrease in the calculated indicator is observed. A statistically significant difference was assessed between the method using ensemble diversity maintenance and standard methods for designing neural networks based on evolutionary algorithms.

Table 2. The results of applying methods with no noise.

Approach	Test Problem 1	Test Problem 2	Test Problem 3	Test Problem 4	Test Problem 5	Concrete Slump Data	Ore-thermal Melting Data
Stochastic Gradient Boosting	0.815	0.976	0.934	0.898	0.862	0.984	0.785
MARSplines	0.785	0.981	0.931	0.826	0.773	0.985	0.813
Single Neural Network	0,928	0.968	0.965	0.947	0.851	0.987	0.806
GASEN	0.987	0.996	0.995	0.983	0.902	0.997	0.934
GA-based	0.981	0.997	0.995	0.981	0.901	0.992	0.941
Proposed approach without diversity maintaining	0.983	0.997	0.996	0.983	0.931	0.994	0.961
Proposed approach with diversity maintaining	0.986	0.996	0.996	0.984	0.951	0.995	0.985

Table 3. The results of applying methods with sample noise of 10%.

Approach	Test Problem 1	Test Problem 2	Test Problem 3	Test Problem 4	Test Problem 5
Stochastic Gradient Boosting	0.701	0.911	0.867	0.746	0.776
MARSplines	0.678	0.924	0.854	0.682	0.680
Single Neural Network	0.764	0.917	0.883	0.785	0.757
GASEN	0.931	0.969	0.935	0.923	0.812
GA-based	0.930	0.958	0.942	0.937	0.784
Proposed approach without diversity maintaining	0.939	0.962	0.947	0.937	0.829
Proposed approach with diversity maintaining	0.952	0.987	0.965	0.946	0.880

The complication of the problem associated with the imposition of 20% noise led to an even greater discrepancy in the effectiveness of individual solvers and solvers built on the basis of groups of neural network models. A statistically significant result allows us to

Table 4. The results of applying methods with sample noise of 20%.

Approach	Test Problem 1	Test Problem 2	Test Problem 3	Test Problem 4	Test Problem 5
Stochastic Gradient Boosting	0.637	0.860	0.821	0.639	0.721
MARSplines	0.628	0.851	0.752	0.641	0.626
Single Neural Network	0.681	0.845	0.819	0.695	0.689
GASEN	0.865	0.889	0.891	0.882	0.739
GA-based	0.850	0.905	0.885	0.920	0.721
Proposed approach without diversity maintaining	0.878	0.904	0.912	0.918	0.795
Proposed approach with diversity maintaining	0.927	0.917	0.933	0.936	0.839

provide exactly the proposed approach, with a modification that ensures the diversity of individual regressors in the ensemble. Given that the available computational resources were equalized across different types of regressors, a statistically significant difference in the results of individual and ensemble solvers seems useful in real-world applications. Since the series of experiments used cross-validation schemes trusted for such studies, this allows us to emphasize the ability of the ensemble approach while maintaining diversity to increase the efficiency of solving complex regression problems, the datasets of which are potentially noisy due to objective circumstances. For the problem under consideration with real ore-thermal smelting data, it was also possible to improve the accuracy of the model. An increase in the accuracy of the model for the problem under consideration with real ore-thermal smelting data was also achieved.

4   Conclusion

The study, the results of which are presented in this article, is aimed at creating and evaluating the effectiveness of new methods for constructing ensemble models based on artificial neural networks. Taking into account the high complexity of constructing the structures of artificial neural networks, the study is focused on automating the process of forming an ensemble. A method for forming the structure of neural networks based on a probabilistic evolutionary procedure is considered and described, which makes it possible to automatically generate the structures of neural network models. This author's approach is complemented by new operations aimed at maintaining diversity in the formed ensemble, which is an important factor in ensuring efficiency in the ensemble approach. Despite the fact that direct estimates of the diversity of the models were not carried out (and this is the subject of a detailed statistical study in the future), the conducted numerical study without the use of appropriate operations shows a statistically significant advantage of the diversity maintenance approach. The relative error reduction is 5–7% depending on the problem.

Comparative studies have shown the advantage or equality of the proposed approach in efficiency over alternative evolutionary methods for forming ensembles of neural network regressors. Moreover, this approach requires tuning a smaller number of parameters compared to the standard evolutionary theory used to design neural network structures. This change does not reduce the ability of the approach to adapt to specific problems, as shown during computational experimentation. For a number of problems, the proposed method made it possible to achieve a relative decrease in the error by about 20%. Another advantage of the approach, leading to a reduction in routine computational operations, is the absence of the requirement for encoding and decoding neural network structures into a binary string, which is typical for standard evolutionary heuristics. We carefully used the ANOVA method to check the correctness of the results obtained, which allowed us to establish cases of truly different results in terms of the efficiency of constructing regression solvers. The implementation of the considered approaches in a software system allows us to continue research in the future and implement applications in solving real data analysis problems. In further studies, it is planned to test and adapt the method for efficient use on larger data sets, the processing of which seems to be a resource-intensive problem for ensemble methods.

References

1. Smith, J.L.: Advances in neural networks and potential for their application to steel metallurgy. Mater. Sci. Technol. **36**(17), 1805–1819 (2020)
2. Guo, L., Lei, Y., Li, N., Yan, T., Li, N.: Machinery health indicator construction based on convolutional neural networks considering trend burr. Neurocomputing **292**, 142–150 (2018)
3. Yamashita, R., Nishio, M., Do, R.K.G., Togashi, K.: Convolutional neural networks: an overview and application in radiology. Insights Imaging **9**(4), 611–629 (2018)
4. AL-Qutami, T.A., Ibrahim, R., Ismail, I., Ishak, M.A.: Virtual multiphase flow metering using diverse neural network ensemble and adaptive simulated annealing. Expert Syst. Appl. **93**, 72–85 (2018)
5. Ribeiro, G.T., Mariani, V.C., dos Santos Coelho, L.: Enhanced ensemble structures using wavelet neural networks applied to short-term load forecasting. Eng. Appl. Artif. Intell. **82**, 272–281 (2019)
6. Melin, P., Monica, J.C., Sanchez, D., Castillo, O.: Multiple ensemble neural network models with fuzzy response aggregation for predicting COVID-19 time series: the case of Mexico. Healthcare **8**(2), 181 (2020)
7. Hansen, L.K., Salamon, P.: Neural network ensembles. IEEE Trans. Pattern Anal. Mach. Intell. **12**(10), 993–1001 (1990)
8. Irvine, N., Nugent, C., Zhang, S., Wang, H., Ng, W.W.: Neural network ensembles for sensor-based human activity recognition within smart environments. Sensors **20**(1), 216 (2019)
9. Li, S., Yao, Y., Hu, J., Liu, G., Yao, X., Hu, J.: An ensemble stacked convolutional neural network model for environmental event sound recognition. Appl. Sci. **8**(7), 1152 (2018)
10. Alzubi, J.A., Bharathikannan, B., Tanwar, S., Manikandan, R., Khanna, A., Thaventhiran, C.: Boosted neural network ensemble classification for lung cancer disease diagnosis. Appl. Soft Comput. **80**, 579–591 (2019)
11. Jia, D.W., Wu, Z.Y.: Seismic fragility analysis of RC frame-shear wall structure under multidimensional performance limit state based on ensemble neural network. Eng. Struct. **246**, 112975 (2021)

12. Li, H., Wang, X., Ding, S.: Research and development of neural network ensembles: a survey. Artif. Intell. Rev. **49**(4), 455–479 (2018)
13. Shu, C., Burn, D.H.: Artificial neural network ensembles and their application in pooled flood frequency analysis. Water Resour. Res. **40**(9) (2004)
14. Giacinto, G., Roli, F.: Design of effective neural network ensembles for image classification purposes. Image Vis. Comput. **19**(9–10), 699–707 (2001)
15. Bukhtoyarov, V.V., Semenkina, O.E.: Comprehensive evolutionary approach for neural network ensemble automatic design. In: Proceeding of the 2010 IEEE World Congress on Computational Intelligence, WCCI 2010—2010 IEEE Congress on Evolutionary Computation, Barcelona, Spain, 18–23 July 2010, p. 5586516 (2010)
16. Bukhtoyarov, V.V., Tynchenko, V.S., Nelyub, V.A., Masich, I.S., Borodulin, A.S., Gantimurov, A.P.: A study on a probabilistic method for designing artificial neural networks for the formation of intelligent technology assemblies with high variability. Electronics **12**, 215 (2023). https://doi.org/10.3390/electronics12010215
17. Khwaja, A.S., Anpalagan, A., Naeem, M., Venkatesh, B.: Joint bagged-boosted artificial neural networks: Using ensemble machine learning to improve short-term electricity load forecasting. Electric Power Syst. Res. **179**, 106080 (2020)
18. Liu, L., et al.: Deep neural network ensembles against deception: ensemble diversity, accuracy and robustness. In: 2019 IEEE 16th International Conference on Mobile Ad Hoc and Sensor Systems (MASS), pp. 274–282. IEEE (2019)
19. Ai, S., Chakravorty, A., Rong, C.: Household power demand prediction using evolutionary ensemble neural network pool with multiple network structures. Sensors **19**(3), 721 (2019)
20. Huang, C., Li, M., Wang, D.: Stochastic configuration network ensembles with selective base models. Neural Netw. **137**, 106–118 (2021)
21. Van Roode, S., Ruiz-Aguilar, J.J., González-Enrique, J., Turias, I.J.: An artificial neural network ensemble approach to generate air pollution maps. Environ. Monit. Assess. **191**(12), 1–15 (2019)
22. Ahvanooey, M.T., Li, Q., Wu, M., Wang, S.: A survey of genetic programming and its applications. KSII Trans. Internet Inf. Syst. **13**(4), 1765–1794 (2019)
23. UCI machine learning repository. https://ergodicity.net/2013/07/. Accessed 21 Feb 2023
24. Yeh, I.C.: Modeling slump of concrete with fly ash and superplasticizer. Comput. Concrete Int. J. **5**(6), 559–572 (2008)
25. Gackowski, M., Szewczyk-Golec, K., Pluskota, R., Koba, M., Mądra-Gackowska, K., Woźniak, A.: Application of multivariate adaptive regression splines (MARSplines) for predicting antitumor activity of anthrapyrazole derivatives. Int. J. Mol. Sci. **23**(9), 5132 (2022)
26. Qin, W., Wang, L., Liu, Y., Xu, C.: Energy consumption estimation of the electric bus based on grey wolf optimization algorithm and support vector machine regression. Sustainability **13**(9), 4689 (2021)
27. Friedman, J.H.: Stochastic gradient boosting. Comput. Stat. Data Anal. **38**(4), 367–378 (2002)
28. Lee, M.C., et al.: Computer-aided diagnosis of pulmonary nodules using a two-step approach for feature selection and classifier ensemble construction. Artif. Intell. Med. **50**(1), 43–53 (2010)
29. Yao, C., Dai, Q., Song, G.: Several novel dynamic ensemble selection algorithms for time series prediction. Neural. Process. Lett. **50**(2), 1789–1829 (2019)
30. Koonce, B., Koonce, B.: EfficientNet. Convolutional Neural Netw. Swift Tensorflow: Image Recogn. Dataset Categorization **1**, 109–123 (2021)
31. Van de Ven, G.M., Siegelmann, H.T., Tolias, A.S.: Brain-inspired replay for continual learning with artificial neural networks. Nat. Commun. **11**, 4069 (2020)
32. Mingxing, T., Quoc, V. Le.: EfficientNetV2: smaller models and faster training. In: International Conference on Machine Learning, vol. 139, pp.1–11 (2021)

33. Gaba, S., Budhiraja, I., Kumar, V., Garg, S., Kaddoum, G., Hassan, M.M.: A federated calibration scheme for convolutional neural networks: models, applications and challenges. Comput. Commun. **192**, 144–162 (2022)
34. Raska, S.: Python and Machine Learning. DMK Press, Moscow (2017)

Self-adaptation Method for Evolutionary Algorithms Based on the Selection Operator

Pavel Sherstnev[✉] [iD]

Artificial Intelligence Laboratory, Siberian Federal University, Krasnoyarsk, Russia
sherstpasha99@gmail.com

Abstract. Genetic algorithms are a class of effective and popular black box optimization methods that are inspired by evolutionary processes in the nature. Genetic algorithms are useful in cases where nothing is known about the optimization object except the inputs and outputs. Such an algorithm iteratively searches for a solution in the solution space based on a predefined fitness function that allows comparing different solutions. If a researcher desires to use genetic algorithms, it becomes necessary to choose genetic operators and numerical parameters of the algorithm, the choice of which may be a difficult task. Self-adaptation methods that alter the behavior of the algorithm while it is running help to deal with the task of choosing the optimal settings of the algorithm. Such methods are called methods of self-adaptation of evolutionary algorithms and are usually divided into self-tuning, which performs the tuning of numerical parameters, and self-configuring, which makes the choice of genetic operators. In recent decades, various strategies for self-adaptation of evolutionary algorithms have been actively developed, including metaheuristic algorithms, as a result of which a researcher can obtain a specialized evolutionary algorithm that solves problems from a certain class better than conventional algorithms. However, even when using the metaheuristic approach, there is a need to choose genetic operators and numerical parameters of the algorithm. Therefore, the subject of the development of self-adaptive algorithms is one of the most relevant fields in the study of evolutionary algorithms. In this paper, a new approach to the adaptation of evolutionary algorithms based on the selection of genetic operators is proposed. The method is applied to the genetic algorithm and compared with the most popular SelfCGA self-configuring approach and shows an improvement in efficiency on both real and binary optimization problems.

Keywords: Self-adaptation · Evolutionary algorithms · Optimization · Genetic algorithm · Selection

1 Introduction

Many practical problems do not have a well-developed analytical solution either because of the frequent appearance of new problems, or because of the absence of a mathematical representation. An example can be any problems from the reinforcement learning class in which there is only a reward function, but there is no mathematically described object that is required by gradient optimization methods. The way out for the researcher in such a situation can be evolutionary algorithms, since for their work it is enough to define the function of mapping the genotype into a phenotype and a function of evaluating the solution (a fitness function). However, in practice, if a researcher wants to use evolutionary algorithms, he also needs some experience working with them, since for any evolutionary algorithm there are many different genetic operators on which the result of optimization depends. The researcher can try all the operators in order to find the best operator for a given problem, but when a new problem appears, this choice may not be optimal, and again it will be necessary to search for a good combination of operators. Therefore, various strategies for self-adaptation of evolutionary algorithms have been developed for a long time [1–3], and sometimes even automatically creating new genetic operators that would provide solutions to problems from a certain class better than conventional ones [4, 5].

However, despite the success of using metaheuristic approaches, the subject of developing methods of self-adaptation of evolutionary algorithms is important, since even in the metaheuristic approach there is a problem of choosing the optimal operators of the evolutionary algorithm, which stands at the top level. Self-adaptation methods are usually divided into two categories: self-configuring methods, which include methods that select the types of operators in the process of the algorithm; self-tuning methods, which are understood as methods for adjusting the real parameters of the algorithm, for example, the probability of mutation. As a rule, each type of evolutionary algorithm has its own set of self-adaptation methods. For example, adaptation algorithms based on the history of successful applications have been developed for the differential evolution method [6–8]. For the genetic algorithm and the genetic programming algorithm, self-configuring methods have been developed that adapt the probabilities of using genetic operators in the process of the algorithm [1] and self-tuning methods that set the behavior of the probability of mutation [2]. The method of self-configuring of evolutionary algorithms described in [1] is the most effective and utilized self-configuring method that can be used to select operators of any evolutionary algorithm. However, this method implies only the choice of operators, but not the numerical parameters of the algorithm.

In this paper, we propose a simple method of self-adaptation of both genetic operators and numerical parameters of an evolutionary algorithm, based on existing selection methods. The method is compared with SelfCGA using the genetic algorithm for solving binary and real-valued optimization problems.

2   Methods of Self-adaptation for Genetic Algorithm

2.1   Genetic Algorithm

Genetic algorithms (GA) are a family of evolutionary optimization algorithms that search for a solution in a hypercube defined by a binary string in which an individual is encoded. The goal of GA is to determine which point of a given hypercube delivers the extremum of the target function. Each of the individuals in the population is one of the points in the whole space of valid solutions. The GA cycle will be described step by step below [9].

The first population is initialized randomly uniformly throughout the search space. Then, using selection, crossover, and mutation operators, the individuals evolve. At different points in the search space, the fitness value of an individual varies.

The selection process, based on the fitness value, selects the individuals that, using crossover operators, will create the offspring for the next generation. It is usually accepted to distinguish the following variants of the selection operator: proportional selection - probability of choosing an individual is proportional to its value of the fitness function; ranked selection - probability of choosing an individual is proportional to the rank of its value of the fitness function; tournament selection - the individual with the best value of the fitness function from a randomly formed subgroup is chosen.

The individuals selected at the previous stage participate in the crossover operation. The purpose of crossover is to provide useful information to the offspring. There are three types of crossover: one-point crossover - parents' chromosomes are randomly divided into two segments, and parents exchange them; two-point crossover - parents' chromosomes are randomly divided into three segments, and parents exchange them; uniform crossover - each bit in the offspring chromosome is inherited randomly from one of the parents.

The final step is mutation. The mutation operator usually makes small random changes in the genotype, thus maintaining diversity in the population. There is only one mutation operator in GA. Only the probabilities with which this mutation is carried out differ. When the mutation operator is over, the GA cycle is repeated again, and selection operators are again applied to the offspring.

If you use GA to work with real or integer variables, each variable must be encoded in a binary form. The easiest way is to divide the number into intervals and make a relation between each interval and the binary combination. In [10] it was found that the most efficient method of encoding real variables into binary ones is Gray code, although in a study [11] it was shown when using a genetic algorithm with varying string length the most efficient way of encoding may be Rice code.

2.2 Fitness-Based Adaptation

A simple method for tuning mutation and crossover probabilities was proposed in [2]. The method is designed to both protect good solutions from destruction and explore the search space by mutating inefficient solutions. Crossover and mutation probabilities are determined for each individual separately and are calculated based on the fitness of those individuals. So, if the fitness of an individual is high, then the probability of crossover and mutation is low (for an individual with the highest fitness these probabilities are 0), but if the fitness of an individual is low, then the probability of crossover and mutation is high. This method has shown to improve the performance on a number of problems of different classes, including the traveling salesman problem and the optimization of neural network weight coefficients. In [12] this method was modified, which improved the performance of the genetic algorithm on the traveling salesman problem. In difference from the original method, for individuals with the highest fitness the probabilities of crossover and mutation are not zero, but are performed with some minimal probability. This increases the speed of convergence to the optimal solution.

2.3 Population-Level Dynamic Probabilities

This method is taken from [3] and is used to tune the probabilities of applying genetic operators (self- configuring). At the start of the algorithm start each genetic operator is assigned minimum probabilities of its application $p_{all} = 0.2/n$, where n is the number of operators of a certain type. Then for each operator the values are entered (formula 1):

$$r_i = \frac{success_i^2}{used_i},$$ (1)

where $used_i$ is the number of uses of the given operator, $success_i$ is the number of successful uses of the operator when the fitness of the offspring exceeded the fitness of the parent. For each operator the probability of its use pi is calculated by the formula 2:

$$p_i = p_{all} + \left[r_i \frac{1.0 - np_{all}}{scale} \right], scale = \sum_{j=1}^{n} r_j.$$ (2)

The $success_i$ values are squared because of very low success rate of genetic operators. The scale values allow to normalize the probability values so that their sum is always equal to one. Among the shortcomings of the *PDP* method it is should be noted that the procedure for evaluating the success of an operator has some disadvantages, in particular, when comparing the succession fitness with that of the parents it is not obvious with which of the parents to compare.

2.4 Self-configuring Genetic Algorithm

The *SelfCEA* method is based on increasing the probability of choosing the operator that provided the highest average fitness on a given generation. Let z_p be the number of different operators of type k. The initial selection probability of each operator is defined as $p_i = 1/z$. The probability of an operator providing the maximum mean value of the fitness function is increased by the formula 3 [1]:

$$p_i = p_i + \frac{(z - 1)K}{zN}, i = 1, 2, ..., z,$$ (3)

where N is the number of generations of algorithm, K is constant controlling rate of change of probability. Probabilities of other operators are recalculated so that the total sum equals 1. The method has two parameters: K, which determines the rate of change in the probability; a threshold which is the minimum probability of using the operators. The method has shown good results in solving the problem of selecting efficient options for the spacecraft control system [13], and has also been modified with new crossover operators in [14] and [15]. In this paper, this method is applied to a genetic algorithm and denoted as *SelfCGA*.

2.5 SHAGA

This algorithm is a successful attempt to apply a strategy for real parameter adaptation of the differential evolution algorithm, called success history based parameter adaptation

(SHA) [7], to the genetic algorithm. To be able to apply this adaptation strategy, the genetic algorithm itself had to undergo several modifications that made the cycle of the genetic algorithm closer to the cycle of differential evolution. Tournament selection with size of two was used as the selection operator. Only one parent was selected for crossover and the second parent was always the current individual in the population. Crossover was performed uniformly, where the crossover probability was tuned similarly to the SHADE algorithm. The method was tested on binary and real-valued optimization problems and achieved better performance than SelfCGA [16].

3 Proposed Approach

The self-configuring approaches discussed above, such as SelfCGA and PDP, perform a selection of genetic operators based on their efficiency during the execution of the algorithm. The operators are first generated with equal probability, and then the probabilities are recalculated in favour of the more efficient ones, depending on the fitness achieved by executing these operators. Ultimately, these self-configuring methods perform the same function: they select genetic operators based on probabilities that depend on the values of the fitness functions.

SelfCGA and PDP algorithms have disadvantages when used to tune a genetic algorithm. SelfCGA has two parameters: the rate of change in probability and a threshold probability that can be assigned to an operator. For different problems and numbers of operators, these parameters must be different. In addition to this, the method is a self-configuring method, which by definition and in practice does not allow for the setting of numerical parameters such as the mutation probability. With SelfCGA, mutation probability has three values (weak, average and strong mutation), which allows you to adjust these parameters, but only to a pre-defined value.

The PDP method, on the other hand, adjusts the probabilities of applying operators not directly on the fitness basis but by creating a progeny with a higher fitness than the parent. This peculiarity creates complications because it is not clear with which parent the offspring should be compared. The method also allows only the genetic operators to be set, but not the numerical values of the algorithm parameters.

Researchers usually develop complex procedures for calculating probabilities based on fitness values, but in evolutionary algorithms there are already functions that assign a selection probability to each fitness function value from the population. These are selection operators.

If individuals were generated using different genetic operators in the previous population, a fitness value can be calculated for each individual, and the information about the type of operator that was used can be used. By having sets of fitness function values and different types of genetic operators that generated individuals with the corresponding fitness values, each time the next generation of individuals is created, the selection operator can be used to select the operator that will produce the offspring.

Let *operator_set* be the set of all genetic operator types of a particular type. N is population size. On the first generation, when the i-th offspring is created, the genetic operator is chosen randomly from this set (formula 4):

$$Operator_i = RandomChoice(operator\_set), \tag{4}$$

where *RandomChoice* is a function for selecting a random item from a set. The probability of each element being selected is equal. *Operator$_i$* is used to create an offspring and *Fitness$_i$* fitness of the offspring is associated with this operator. Once the new generation is fully formed, there is a set of values (formula 5):

$$OperList = \begin{bmatrix} Operator_1 & Fitness_1 \\ \vdots & \vdots \\ Operator_N & Fitness_N \end{bmatrix}. \tag{5}$$

Now, when creating the *i*-th offspring, the operator is selected as follows (formulas 6 and 7):

$$If \ rand \ < p_t: Operator_i = RandomChoice(operator\_set) \tag{6}$$

$$Else: Operator_i = Selection(OperList) \tag{7}$$

where *rand* is a random value generated by a uniform distribution in the range 0 to 1, p_t is the probability with which an operator is randomly selected from the set of possible operator_set variants, *Selection* is a genetic selection operator that selects an individual (in this case another genetic operator from *OperList* based on the values of the fitness function.

The condition in formula 3 ensures the robustness of the adaptation process by ensuring that different operators are present in OperList, creating permanent competition.

Now consider the procedure for adapting a numerical parameter. Let *Left* be the minimum value of the parameter and *Right* be the maximum value of the parameter. N - population size. In the first generation, when the *i*-th offspring is created, the parameter is generated by a uniform distribution (formula 8):

$$Parametr_i = Uniform(Left, Right) \tag{8}$$

where *Uniform* is the function that generates a random value in the *Left* and *Right* boundaries.

Parametr$_i$ is used to create a descendant and fitness *Fitness$_i$* per offspring is associated with this operator. Once the new generation is fully formed, there is a set of values (formula 9):

$$ParamList = \begin{bmatrix} Parametr_1 & Fitness_1 \\ \vdots & \vdots \\ Parametr_N & Fitness_N \end{bmatrix} \tag{9}$$

Now, when creating the *i*-th offspring, the operator is selected as follows (formulas 10 and 11):

$$If \ rand \ < p_t: Parametr_i = Uniform(Left, Right) \tag{10}$$

$$Else: Parametr_i = Selection(ParamList) \tag{11}$$

Thus, the proposed method allows to tune both the numerical parameters of the algorithm and the genetic operators, which makes it a self-adaptive method. In future work, this method will be referred to as SelfAGA. It is easy to see that there can be as many variations of this method as there are selection functions and considering that a selection function can be generated using metaheuristics, this method can be used for automated generation of self-adaptive evolutionary algorithms.

In the next section, we compare the proposed method with SelfCEA self-configuring method for solving binary and real-valued optimization problems by genetic algorithm.

4   Experimental Setup and Results

4.1   Parameters of the Genetic Algorithm

A genetic algorithm was implemented in the python programming language and then modified to match each method. The SelfCGA and SelfAGA algorithms use the same implementations of genetic operators and the other genetic algorithm process is programmed in the same way, except for the self-adaptation method. Thus the difference in algorithm performance can only be due to different adaptation methods. Table 1 below shows the algorithm parameters and operators involved in adaptation.

Table 1. Parameters of the genetic algorithm and self-adaptation methods

Name of the parameter	Value
type of crossover	one-point, two-point, uniform
type of selection	proportional, rank, tournament (5)
type of mutation (SelfCGA)	weak, average, strong
K (SelfCGA)	0.5
threshold (SelfCGA)	0.05
P_t (SelfAGA)	0.1
Selection (SelfAGA)	proportional, rank, tournament (3 and 5)

The average mutation performs a bit flip with a probability that is calculated as follows (formula 12):

$$p = \frac{1}{StringLen} \tag{12}$$

where *StringLen* is the length of the binary string. Weak and strong mutations are three times less and three times more likely, respectively.

Since in the SelfAGA method the probability is adjusted as a real parameter, the minimum value is the probability determined in the weak SelfCGA mutation, and the maximum value is the probability determined in the strong mutation. The remaining part of the paper is a comparison of SelfCGA with SelfAGA, with SelfAGA presented in four versions, distinguished by the selection function used for self-adapted parameters (proportional, rank and tournament with size of 3 and 5).

4.2 Results of Solving Binary Optimization Problems

A total of two binary problems are used: the problem "onemax", which contains one global optimum and consists in obtaining the maximum number of 1 in the binary string. The optimum in this problem is reached when all elements of the binary string are 1.

Another problem is called "01". The goal is to find the binary string with the maximum number of pairs 0 and 1. Unlike "onemax", this problem has one global and one local optimum. For both problems "onemax" and "01" there are different variations, varying in the dimensionality of the problem, as well as the number of generations and population size. Table 2 below shows the parameters of the problems.

Table 2. Parameters of binary optimization problems

Problem	Number of iterations	Population size
Onemax 1000 bits	100	100
Onemax 3000 bits	100	100
Onemax 10000 bits	1000	100
«01» 1000 bits	100	100
«01» 3000 bits	100	100

Table 3 below shows the best solutions achieved by each of the self-adaptation methods, averaged over 1000 independent runs.

Table 3. Results of solving binary optimization problems.

Heading level	SelfCGA	SelfAGA (proportional)	SelfAGA (rank)	SelfAGA (tournament 3)	SelfAGA (tournament 5)
Onemax 1000 bits	891.075	**899.699**	**900.404**	**905.372**	**914.516**
Onemax 3000 bits	2204.902	**2211.933**	2207.496	**2227.298**	**2264.675**
Onemax 10000 bits	8444.079	**8601.686**	**8562.07**	**8583.36**	**8662.617**
«01» 1000 bits	402.348	**405.309**	**405.238**	**404.676**	**404.942**
«01» 3000 bits	1035.684	1035.135	1034.647	1035.198	**1038.67**

For statistical verification, the Mann-Whitney test is used with a significance level of 0.01. In the case of Null hypothesis there are no differences between observations. Table 3 shows the values in bold for the methods whose difference between the results was statistically significant compared to the SelfCGA results.

4.3 Results of Solving Real-Valued Optimization Problems

The second part of the experiments consists of solving real optimization problems, namely 10 functions from [17]. The number of iterations and the size of the iteration were chosen for each problem individually. Table 4 below shows the functions from [17] as well as the parameters under which the functions have been optimized. Only a part of the original set of functions is used due to the fact that the genetic algorithm could not find an optimal solution for a limited number of iterations (the test calculations were performed up to a number of iterations of 1000 and a generation size of 1000).

Table 4. Parameters of real-valued optimization problems.

Problem	Number of iterations	Population size
F1	30	30
F2	45	45
F4	45	45
F5	100	50
F6	1000	500
F9	100	50
F10	300	200
F12	70	50
F15	100	50
F16	1000	500

The reliability criterion for the genetic algorithm is the ratio of optimal solutions found out of all runs of the algorithm (reliability). For genotype-phenotype mapping a string of bits to real values Gray code and 16 bits per variable were used. Table 5 below shows the average reliability of algorithms averaged over 1000 runs.

The highest values achieved on a given problem are shown in bold. It can be noticed that the average reliability for all the problems solved by the new self-adaptation method surpassed the reliability achieved by the SelfCGA method. The maximum average reliability was achieved using rank selection, while using proportional selection did not significantly improve the reliability of the genetic algorithm on the given tasks. Consider as an example the process of adaptation of the genetic algorithm parameters on the F5 problem. The results are averaged over 1000 runs. Figure 1 below shows the number of various crossover operators (upper plot), selection (middle plot) and the average value of mutation probability (lower plot).

The graph shows how these values adapt over the running of the algorithm.

Table 5. Results of solving real-valued optimization problems.

Problem	SelfCGA	SelfAGA (proportional)	SelfAGA (rank)	SelfAGA (tournament 3)	SelfAGA (tournament 5)
F1	0.740	0.658	0.913	0.927	**0.939**
F2	0.757	0.801	0.878	**0.883**	0.864
F4	0.763	0.762	**0.834**	0.829	0.805
F5	0.621	0.468	**0.637**	0.601	0.568
F6	0.744	**0.988**	0.799	0.791	0.759
F9	0.740	**0.751**	0.697	0.681	0.666
F10	0.690	0.694	0.785	0.769	**0.787**
F12	0.630	0.548	**0.747**	0.732	0.724
F15	0.726	**0.735**	0.701	0.680	0.679
F16	0.554	0.610	0.834	0.878	**0.914**
mean	0.6965	0.7015	**0.7825**	0.7771	0.7705

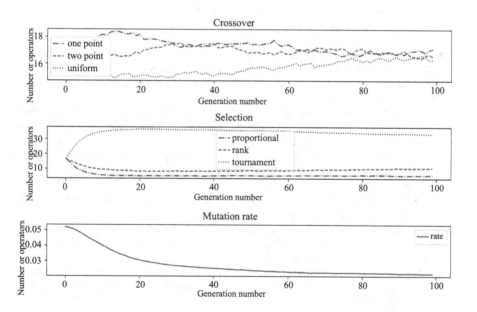

Fig. 1. The process of adaptation of the parameters of the genetic algorithm

5 Conclusion

In this paper, a method for self-adaptation of evolutionary algorithms is proposed and tested that differs from the known methods of parameter selection and the possibility of tuning both genetic operators and numerical parameters of the algorithm. The proposed algorithm can be presented in different variants depending on the selection function that performs algorithm parameter adaptation in the process. The method is compared with the popular SelfCGA approach and shows better performance on both binary and real-valued optimization problems. Further work will be done to study the effect of pt on the efficiency of the method and investigate the performance of the algorithm as compared to other self-adaptation methods on a wider set of problems.

Acknowledgments. This work was supported by the Ministry of Science and Higher Education of the Russian Federation (Grant № 075-15-2022-1121).

References

1. Semenkina, M.: Self-adaptive evolutionary algorithms for designing information technologies for data mining. Artif. Intell. Decis.-Making **1**, 12–24 (2012)
2. Patnaik, L.: Adaptive probabilities of crossover and mutation in genetic algorithms. IEEE Trans. Syst. Man Cybern. **24**, 656–667 (1994)
3. Niehaus, J., Banzhaf, W.: Adaption of operator probabilities in genetic programming. In: Miller, J., Tomassini, M., Lanzi, P.L., Ryan, C., Tettamanzi, A.G.B., Langdon, W.B. (eds.) EuroGP 2001. LNCS, vol. 2038, pp. 325–336. Springer, Heidelberg (2001). https://doi.org/10.1007/3-540-45355-5_26
4. Richter, S.: The automated design of probabilistic selection methods for evolutionary algorithms. In: Proceedings of the 2018 Genetic and Evolutionary Computation Conference Companion, pp. 1545–1552 (2018)
5. Hong, L., Woodward, J., Özcan, E., Liu, F.: Hyper-heuristic approach: automatically designing adaptive mutation operators for evolutionary programming. Complex Intell. Syst. **7**, 3135–3163 (2021). https://doi.org/10.1007/s40747-021-00507-6
6. Zhang, J., Sanderson, A.C.: JADE: adaptive differential evolution with optional external archive. IEEE Trans. Evol. Comput. **13**(5), 945–958 (2009)
7. Tanabe, R., Fukunaga, A.: Success-history based parameter adaptation for differential evolution. In: Proceedings of 2013 IEEE Congress on Evolutionary Computation, pp. 71–78 (2013). https://doi.org/10.1109/CEC.2013.6557555
8. Tanabe, R., Fukunaga, A.: Improving the search performance of SHADE using linear population size reduction. In: Proceedings of the 2014 IEEE Congress on Evolutionary Computation, pp. 1658–1665 (2014)
9. Holland, J.: Adaptation in Natural and Artificial Systems. MIT Press, Cambridge (1975)
10. Whitley, L.D.: Free lunch proof for gray versus binary encoding. In: Genetic and Evolutionary Computation Conference (1999)
11. Panfilov, I.: Study of the performance of a genetic optimization algorithm with alternative solution representation. Siberian Aerosp. J. **4**(50), 68–71 (2013)
12. Han, S., Xiao, L.: An improved adaptive genetic algorithm. In: SHS Web of Conferences, vol. 140, pp. 5–6 (2022)

13. Semenkin, E., Semenkina, M.: Spacecrafts' control systems effective variants choice with self-configuring genetic algorithm. In: Proceedings of 9th International Conference on Informatics in Control, Automation and Robotics, pp. 84–93 (2012)
14. Semenkin, E., Semenkina, M.: Self-configuring genetic programming algorithm with modified uniform crossover. In: Proceedings of 2012 IEEE Congress on Evolutionary Computation, CEC 2012 (2012)
15. Semenkin, E., Semenkina, M.: Self-configuring genetic algorithm with modified uniform crossover operator. In: Tan, Y., Shi, Y., Ji, Z. (eds.) ICSI 2012. LNCS, vol. 7331, pp. 414–421. Springer, Heidelberg (2012). https://doi.org/10.1007/978-3-642-30976-2_50
16. Stanovov, V., Akhmedova, S., Semenkin, E.: Genetic algorithm with success history based parameter adaptation. In: Proceedings of 11th International Conference on Evolutionary Computation Theory and Applications, pp. 180–187 (2019)
17. Suganthan, P.N., et al.: Problem definitions and evaluation criteria. In: Proceedings of CEC 2005 Special Session on Real-Parameter Optimization (2005)

Application of U-Net Architecture Neural Network for Segmentation of Brain Cell Images Stained with Trypan Blue

Vadim Tynchenko$^{(\boxtimes)}$, Denis Sukhanov , Aleksei Kudryavtsev ,
Vladimir Nelyub , Aleksei Borodulin , and Daniel Ageev

Bauman Moscow State Technical University, Moscow, Russia
vadimond@mail.ru

Abstract. This article discusses the problem of semantic segmentation of images of rat brain cells stained with trypan blue. The purpose of this work is to develop software that will automatically perform the segmentation of living brain cells. The originality of this article is a non-standard approach to modifying the architecture of the U-net, which is used to detect brain cells. To solve the problem a mathematical model was developed in the form of a convolutional neural network based on the U-Net architecture with a convolution depth of 4. The initial sample consisted of 30 images, data augmentation was performed, which made it possible to increase the number of examples up to 150 samples. During augmentation images were rotated by 90 degrees clockwise and counterclockwise, and flipped along the X-axis and along the Y-axis. The effect of adding DropOut layers with a probability of 0.3 and/or BatchNormalization on learning and obtaining predictions was considered. When training a deep convolutional neural network, optimizers Adam, SGD, RMSprop were used. The result of applying the selected model according to metric Dice is 0.8513, according to the accuracy of detecting the number of living neurons is 98.62%. A possible future improvement of the current models due to small changes in the architecture of the convolutional neural network, the selection of hyperparameters for optimizers, an increase in the initial samples, and a change in the approach to image preprocessing are considered.

Keywords: Brain Cells · Biomedical Image Segmentation · Image Recognition · Image Preprocessing · Deep Convolutional Networks · U-Net

1 Introduction

Initial bright-field image of brain cells is shown in the Fig. 1. Alternative approach of detecting of live/dead (D/A test) brain cells is using trypan blue staining of brain cells, to develop a mathematical model for the semantic segmentation of images of rat brain cells stained with trypan blue (Fig. 2). The segmentation model will make it possible to recognize living cells automatically.

The relevance of the work is due to the need for a quick and accurate assessment of the viability of brain cells to study the effects of drugs, physical or chemical stimulants, and other potential factors affecting their functioning. Existing methods for assessing viability require the addition of reagents to the substrate with cells for their staining. So, trypan blue staining allows you to identify dead cells, penetrating through their damaged membrane. The developed model of segmentation of images of brain cells will allow tracking the dynamics of cell death in real time. In the future, the model will make it possible to establish the type of cell death, as well as to study the mechanisms that trigger neuronal apoptosis, a key factor in the pathogenesis of diseases and damage to the central nervous system.

The development of the mathematical model was performed at the Scientific and Educational Center "Artificial Intelligence Technologies" (SEC "AI Technologies") as part of the strategic project "Bauman DeepTech", track "Biotechnologies and soft matter". Images of brain cells are provided by the Brain Institute, Moscow. The SEC "AI Technologies" conducts fundamental research in the field of artificial intelligence algorithms, our colleagues have done a number of works in this area [1–4].

Semantic segmentation of images is the split of an image into separate groups of pixels, areas corresponding to one object with simultaneous determination of the type of object in each area. To solve the problem of semantic segmentation, deep convolutional neural networks are often used [5, 6]. The most popular among them are such architectures as AlexNet [7], VGG-16 [8], ResNet-50 [9], U-Net [10], FastFCN [11], Gated-SCNN [12], DeepLab [13], Mask R-CNN [14], W-Net [15]. U-Net and W-Net are used in materials science, in particular, in the metallographic analysis of steel samples [16]. In biomedical tasks, U-Net shows the best results, its unique is able to work with a small number of training images and at the same time gives a high result.

There are various methods such as U-shaped FCN [17] for segmenting neuronal cell bodies in microscopy images and for assessing neurotoxicity from morphological changes in cells observed in different fluorescence modes [18]. There are also methods for determining cell viability using dye-free phase contrast [19] and solving the problem of object detection using Inception [20]. However, all of them are not sufficiently effective for determining the survival of neurons in complex images containing glial cells when stained with trypan blue. Therefore, we decided to use U-Net convolutional neural network to detect live neurons.

Trypan blue staining of brain cells is widely used in biomedicine [21]. This method, due to toxicity, kills cells, which complicates further studies on survival, therefore, it is planned to use bright-field images of cells that are formed under the influence of a direct light beam in the future (Fig. 1).

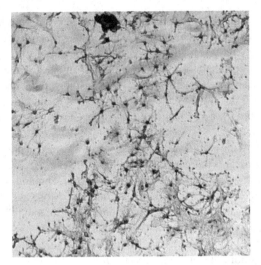

Fig. 1. Bright-field image of rat brain cells.

2   Materials and Methods

Detection of live/dead (D/A test) brain cells in images stained with trypan blue was performed using semantic segmentation by an artificial neural network. The data consisted of 30 RGB images sized 1376 × 1032 pixels, one of the images is shown in (Fig. 2). For each such image, there was a copy of it with live cells marked in red, see (Fig. 3).

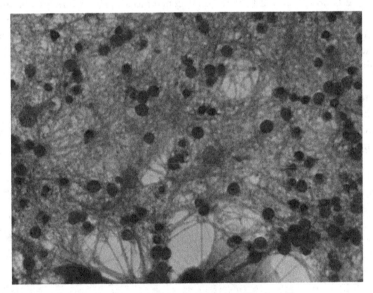

Fig. 2. Trypan blue stained image of brain cells.

To solve the problem of semantic segmentation, a convolutional neural network was used. The teacher for the convolutional neural network is the mask of the original image. Accordingly, for each original image, using markers for living cells, masks of the original images were prepared (Fig. 4). The original images and their prepared masks had a size of 1376 × 1032 pixels, which is not suitable for training a convolutional neural network, so they were compressed to a size of 512 × 512 pixels.

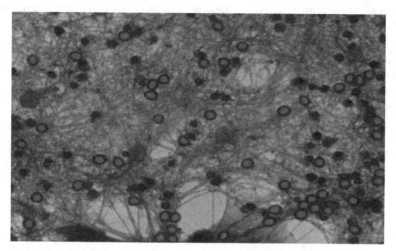

Fig. 3. Image of brain cells with live neurons highlighted.

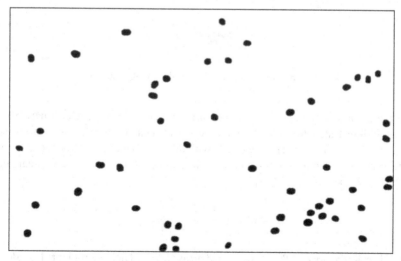

Fig. 4. The mask for the image in Fig. 1 using the markers in Fig. 2.

The architecture of the U-Net network with a depth of 4 was taken as a basis; a schematic representation is shown in Fig. 5. Network inputs are 512 × 512 pixels

RGB images, which passes through 4 encoding blocks. Each encoding block consists of a double Conv2d convolution layer (kernel size 3 × 3, 32 input channels) and a ReLU activation layer and one MaxPool2d layer (2 × 2 kernel) - a pooling operation for downscaling. At each stage of convolution, the number of channels is doubled. Next comes one double convolution block without MaxPool2d and 4 decoding blocks, consisting of a verification layer that reduces the number of channels by 2 times - ConvTranspose2d, two Conv2d convolution layers (3 × 3 kernel) and an activation ReLU layer. At the output, we have a sigmoid activation function and grayscale image of 512 × 512 pixels.

The input data consisted of 30 images, which of course is not enough to get a good result when training a neural network, so image augmentation was done. Each image was augmented 4 times: rotated clockwise by 90°, rotated counterclockwise by 90°, flipped along the X-axis, flipped along the Y-axis. Thus, instead of 30 initial images, we have prepared dataset of 150 samples of input data that we can work with it.

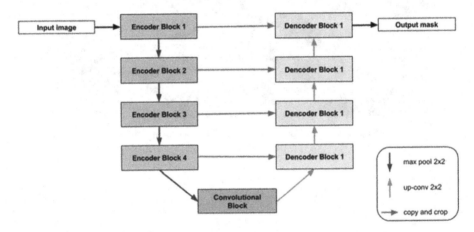

Fig. 5. Blocks of the CNN, U-Net architecture.

To the main architecture of the U-Net artificial convolutional neural network, the effect of adding BatchNormalization2d and/or DropOut (probability is 0.3) layers on the final result was studied. Three variants with different parameters were considered as an optimizer for training: SGD, Adam [22], RMSprop [23]. The metric was calculated using the Dice coefficient (1) [24]:

$$DSC = \frac{2\left|Y^{pred} \cap Y\right| + 1}{\left|Y^{pred}\right| + |Y| + 1} \tag{1}$$

where Y is original array, Y^{pred} is predicted array, DSC – Dice coefficient, 1 is added to avoid possible division by zero when Y^{pred} and Y are arrays of zeros. Loss Function – Combining Dice Losses and Binary Cross Entropy [25] ($BCE_{Dice}loss$), shown in Eq. (2):

$$BCE_{Dice}loss = -\frac{1}{N} \sum_{n=1}^{N} \left[Y_n \log Y_n^{pred} + (1 - Y_n) \log(1 - Y_n^{pred}) \right] + DSC \tag{2}$$

where Y_n is element of array Y, Y_n^{pred} is element of array Y^{pred}, N – number of elements in the array. Experiments, data preprocessing and model training were performed in Python, in particular, the PyTorch framework was used to create the architecture of the deep convolutional neural network. The training of U-Net was performed on a computing machine with two GPUs Nvidia A100 40 Gb.

3   Results

The results of the experiments are presented in Table 1. Abbreviations:

- BN – adding layers BatchNormalization2d;
- DO – adding layers DropOut;
- m – momentum [26];
- w_decay – weight_decay.

Table 1. Experimental results of training CNN models.

Dataset	Type of CNN	CNN depth	Optimazer	Learning rate	Metrics, Dice	Add. Parameters
30	U-Net	4	Adam	1e−3	0.7031	
150	U-Net	4	Adam	1e−3	0.8391	
150	U-Net+BN	4	Adam	1e−3	0.8505	
150	U-Net+DO	4	Adam	1e−3	0.8371	
150	U-Net+BN+DO	4	Adam	1e−3	0.8513	
150	U-Net	4	RMSprop	1e−4	0.7699	
150	U-Net+BN	4	RMSprop	1e−3	0.8421	
150	U-Net+DO	4	RMSprop	1e−4	0.7634	
150	U-Net+BN+DO	4	RMSprop	1e−3	0.8393	
150	U-Net	4	SGD	1e−1	0.8277	m = 0.9, w_decay = 1e−5
150	U-Net+BN	4	SGD	1e−1	0.8239	Nesterov, m = 0.1, w_decay = 1e−5
150	U-Net+DO	4	SGD	1e−1	0.8261	m = 0.8, w_decay = 1e−5
150	U-Net+BN+DO	4	SGD	1e−1	0.8008	Nesterov, m = 0.1, w_decay = 1e−5

The values of the loss function during training models with the Adam optimizer on various architectures of the convolutional neural network are presented in Figs. 6. Dataset was split into 80% of train data, and 20% of test data.

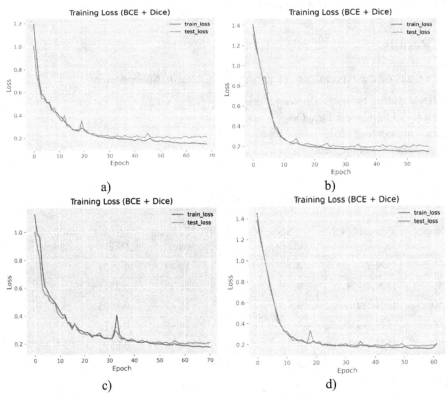

Fig. 6. Loss function training of various models on augmented data: a) U-Net model; b) of U-Net + BN model; c) U-Net + DO model; d) U-Net + BN + DO model.

Early stopping was used while training models. If the loss function has not decreased for 10 epochs in a row, then we stop training and save the model with the best results. The results of prediction by U-Net + BN + DO (Adam) model is shown in Fig. 7. The image shown in Fig. 1 was used to obtain the predicted mask.

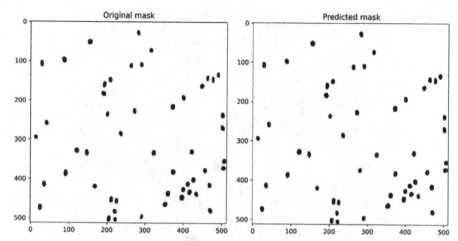

Fig. 7. Comparison of the original mask (left) and the mask predicted by the neural network (right).

4 Discussion

Initial dataset contains 30 images and their masks. Small size of dataset does not allow to get a good result, so data augmentation was implemented first. On the augmented data, without changing other parameters, we see a significant improvement in the metric from 0.7031 to 0.8391 (Table 1). A further change in the architecture of the convolutional neural network, namely the addition of batch normalization and dropout layers, obtain the metric increased to 0.8513. It is important to note that the Adam optimizer is the easiest to use, unlike other optimizers such as SGD and RMSprop, which require selection of parameters depending on the architecture of the U-Net and task type. Figures 6c and 6d show how the dropout layers influence on overfitting by increasing the total number of epochs to an early stopping. The loss curve on the train set and the loss curve on the test set do not diverge too much. In Fig. 7 the predicted mask looks very similar to the original, with the exception of one brain cell, which the neural network has identified as alive. Comparing the original mask and predicted mask, it is important to note that the total number of detected living brain cells matches or in some cases differs by 1–2 brain.

In work [20], more specifically in experiment C, the problem of searching for living cells is also solved, but the Inception neural network is used, which solves the problem of detecting an object in an image. The metric they obtained, precision = 0.92, shows high accuracy, but in our case, it was impossible to solve the problem of "object detection" due to the large number of cells and their clumps in the image, as shown in Fig. 8, as well as the small number of original images. Based on the above conditions, our task cannot be considered as an "object detection" as part of computer vision.

It is important to note that this image was very difficult for visual recognition by a specialist neuroscientist, however, our neural network model coped perfectly with the task of segmenting alive neurons densely located to each other.

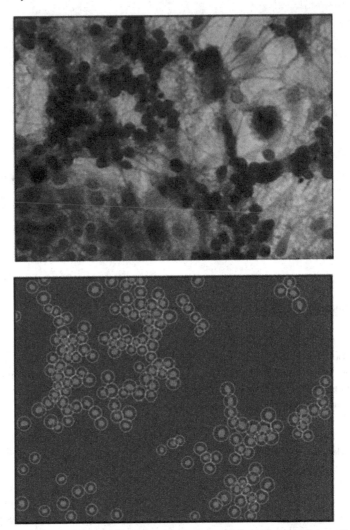

Fig. 8. Trypan blue stained and predicted images with large number of brain cells and their clumps. The number of detected and highlighted cells in the figure is 172.

5   Conclusions

This article is an intermediate step (segmentation of trypan blue staining) in achieving the final purpose (segmentation of bright-field). A good result has been achieved, obtained values were shown in Table 1 and Fig. 8. Among the 13 trained models, the best set is U-Net with batch normalization layers and dropout layers, with 4 convolution depth, Adam optimization, learning rate 0.001. At the moment the best metric (Dice) is 0.8513. To future work, it is proposed to increase the depth of the convolutional neural network. Adjusting of hyperparameters for the Adam optimizer by grid search, random search, or using a genetic algorithm (on binary strings with elitism) and Bayesian optimization (the

Tree-Structured Parzen Estimator method [27]). Similar adjusting of hyperparameters for other optimizers. Consideration of other architectures of neural networks, in particular W-Net. New images of rat brain cells stained with trypan blue are also expected, which will certainly improve model recognition metrics and make it more flexible. Other methods of augmentation will be considered, in addition to those that were present in this article. Moreover, a different approach to image segmentation is assumed. Other approach is not compressing the original image from 1376×1032 pixels to 512×512, but in splitting the original image into samples of 512×512 pixels, thus increasing the dataset.

In addition, in the process of implementing the trained convolutional neural network model in production, in the cytology laboratory, the speed of image processing on the CPU of an ordinary computer (i.e., inference time) will be an important factor. Thus, we need to find a balance between the complexity of the model (inference time directly depends on the number of parameters of the convolutional neural network) and the accuracy of segmentation. This formulation of the problem can be defined as a two-criterion optimization problem. Further work will be devoted to solving this two-criterion optimization problem: minimizing the number of model parameters and maximizing segmentation accuracy by the Dice coefficient. It is assumed to use the "Nondominated Sorting Genetic Algorithm II" (NSGA-II) [28], which is a well-known, fast and elitist multi-objective genetic algorithm. As a result of the NSGA-II algorithm, a Pareto set will be formed, from which the optimal solution will be selected in terms of the balance between image processing speed and segmentation accuracy, based on the priority of factors.

In the future, difficulties are expected when using the U-Net neural network model for images that differ in color. Some images of brain cells may differ from the total mass of the training dataset so much that the neural network model will demonstrate low accuracy metrics of segmentation of alive neurons on them. Neuroscientists point to the reasons for obtaining images that are diverse in color: the use of different microscopes for photofixation, as well as an abnormal amount of trypan blue dye used. As a solution to this problem, it is proposed to consider several methods of stein-normalization of images.

The task of segmentation of alive brain cells in a bright-field image is more complex relative to a similar task in images stained with trypan blue. In this regard, it is planned to develop a methodology for experiments on photofixation of brain cells before and after staining, so that images of alive neurons when superimposed form a pixel-by-pixel correspondence. It is planned that the permissible deviation in pixels is no more than 5% of the total size of the neuron (taking into account the size of the neuron in 20–30 pixels in diameter). The correlation of two images (bright-field and stained with trypan blue) will be performed based on the reference point in the form of a label on the plate. This approach will allow to train a convolutional neural network to segment alive neurons in a bright-field image using segmentation masks obtained from images of brain neurons stained with trypan blue. The result of the work will be a model of identifying and classifying neurons on bright-field images with high accuracy.

References

1. Barantsov, I.A., Pnev, A.B., Koshelev, K.I., Tynchenko, V.S., Nelyub, V.A., Borodulin, A.S.: Classification of acoustic influences registered with phase-sensitive OTDR using pattern recognition methods. Sensors **23**(2), 582 (2023)

2. Bukhtoyarov, V.V., Tynchenko, V.S., Nelyub, V.A., Masich, I.S., Borodulin, A.S., Gantimurov, A.P.: A study on a probabilistic method for designing artificial neural networks for the formation of intelligent technology assemblies with high variability. Electronics **12**(1), 215 (2023)

3. Masich, I.S., et al.: Prediction of critical filling of a storage area network by machine learning methods. Electronics **11**(24), 4150 (2022)

4. Mikhalev, A.S., et al.: The orb-weaving spider algorithm for training of recurrent neural networks. Symmetry **14**(10), 2036 (2022)

5. Khairandish, M.O., Sharma, M., Jain, V., Chatterjee, J.M., Jhanjhi, N.Z.: A hybrid CNN-SVM threshold segmentation approach for tumor detection and classification of MRI brain images. Irbm **43**(4), 290–299 (2022)

6. Van Valen, D.A., et al : Covert: deep learning automates the quantitative analysis of individual cells in live-cell imaging experiments. PLOS Comput. Biol. **12**(11) (2016)

7. Krizhevsky, A., Sutskever, I., Hinton, G.E.: ImageNet classification with deep convolutional neural networks. In: Advances in Neural Information Processing Systems, vol. 25 (2012)

8. Simonyan, K., Zisserman, A.: Very deep convolutional networks for large-scale image recognition. Preprint on subject of Computer Vision and Pattern Recognition. https://arxiv.org/abs/1409.1556. Accessed 10 Apr 2015

9. He, K., Zhang, X., Ren, Sh., Sun, J.: Deep residual learning for image recognition. Tech report on subject of Computer Vision and Pattern Recognition. https://arxiv.org/abs/1512.03385. Accessed 10 Dec 2015

10. Ronneberger, O., Fischer, P., Brox, T.: U-net: convolutional networks for biomedical image segmentation. In: Navab, N., Hornegger, J., Wells, W.M., Frangi, A.F. (eds.) MICCAI 2015. LNCS, vol. 9351, pp. 234–241. Springer, Cham (2015). https://doi.org/10.1007/978-3-319-24574-4_28

11. Wang, Zh., Ji, Sh. : Smoothed dilated convolutions for improved dense prediction. In: KDD 2018: Proceedings of the 24th ACM SIGKDD International Conference on Knowledge Discovery & Data Mining, pp. 2486–2495 (2018). https://doi.org/10.1145/3219819.3219944. Accessed 01 May 2019

12. Takikawa, T., Acuna, D., Jampani, V., Fidler, S.: Gated-SCNN: gated shape CNNs for semantic segmentation. https://nv-tlabs.github.io/GSCNN/. Accessed 12 July 2019

13. Chen, L.-Ch., Papendreou, G., Kokkinos, I., Murphym K., Yullie, A.L.: DeepLab: semantic image segmentation with deep convolutional nets, atrous convolution, and fully connected CRFs. IEEE Trans. Pattern Anal. Mach. Intell. **40**(4), 834–848. https://arxiv.org/abs/1606.00915. Accessed 12 May 2017

14. He, K., Gkiozari, G., Dollar, P., Girshhick, R.: Mask R-CNN. https://arxiv.org/abs/1703.06870. Accessed 24 Jan 2018

15. Xia, X., Kulis, B. W-Net: a deep model for fully unsupervised image segmentation. https://arxiv.org/abs/1711.08506. Accessed 22 Nov 2017

16. Kovun, V.A., Kashirina, I.L.: Usage of U-Net and W-net neural network architectures for steel samples metallographic analysis. Proc. VSU Ser. Syst. Anal. Inf. Technol. **1**, 101–110 (2022)

17. Hu, T., Xu, X., Chen, S., Liu, Q.: Accurate neuronal soma segmentation using 3D multi-task learning U-shaped fully convolutional neural networks. Front. Neuroanat. **14**, 592806 (2021)

18. Wang, S., Linsley, J.W., Linsley, D.A., Lamstein, J., Finkbeiner, S.: Fluorescently labeled nuclear morphology is highly informative of neurotoxicity. Front. Toxicol. **4**, 935438 (2022)
19. Hu, C., et al.: Live-dead assay on unlabeled cells using phase imaging with computational specificity. Nat. Commun. **13**(1), 713 (2022)
20. Christiansen, E.M., et al.: In silico labeling: predicting fluorescent labels in unlabeled images. Cell **173**(3), 792–803 (2018)
21. Melnikova, N.A., Shubina, O.S., Dudenkova, N.A., Lapshina, M.V., Liferenko, O.V., Timoshkina, O.I.: A study of the viability of cells when exposed to lead acetate on the organism of rats. Mod. Probl. Sci. Educ. **5**, 494 (2013)
22. Kingma, D.P., Ba, J.: Adam: a method for stochastic optimization. In: Proceedings of 3rd International Conference for Learning Representations (2014). https://arxiv.org/abs/1412.6980. Accessed 30 Jan 2017
23. Dauphin, Y., Harm D.V., Yoshua B.: Equilibrated adaptive learning rates for non-convex optimization. In: Advances in Neural Information Processing Systems, vol. 28 (2015)
24. Dice, L.R.: Measures of the amount of ecologic association between species. Ecology **26**(3), 297–302 (1945)
25. Murphy, K.: Machine Learning: A Probabilistic Perspective. Massachusetts Institute of Technology (MIT) (2012)
26. Sutskever, I., Martens, J., Dahl, G., Hinton, G.E., Sanjoy, D.: On the importance of initialization and momentum in deep learning. In: Proceedings of the 30th International Conference on Machine Learning (ICML-13), pp. 1139–1147 (2016)
27. Bergstra, J., Rémi, B., Yoshua, B., Balázs, K.: Algorithms for hyper-parameter optimization. In: Advances in Neural Information Processing Systems, vol. 24 (2011)
28. Deb, K., Pratap, A., Agarwal, S., Meyarivan, T.A.M.T.: A fast and elitist multiobjective genetic algorithm: NSGA-II. IEEE Trans. Evol. Comput. **6**(2), 182–197 (2002)

Language Model Architecture Based on the Syntactic Graph of Analyzed Text

Roman Semenov[✉] [ID]

MIREA – Russian Technological University, Vernadsky Avenue 78, 119454 Moscow, Russia
9629790@gmail.com

Abstract. The methods and techniques of graph structures for text processing are considered. The task of processing Russian-language text and extracting semantic structures is an important stage in the development of artificial intelligence systems. Existing models of intelligent assistants are unable to handle a large volume of noisy information and take a long time to process requests. To solve this problem, the article proposes methods for working with graph structures for the analysis and classification of necessary data. By performing initial processing according to the proposed conceptual structure, it becomes possible to use a syntactic graph for a more accurate representation of each part of speech in the processed context. The results of the tested model provided data on the accuracy of word identification in Russian-language sentences. A table comparing the accuracy with existing natural language processing models is presented. The results were obtained based on the fact that 70% of the text volume is required for the training set, and analysis was conducted on the remaining portion, which is true for each of the compared model.

Keywords: Syntactic Graph · Conceptual Structure · Graphlet · Logical Structure · Actionable Construct · Analyser · Information Noise

1 Introduction

Abstract models are becoming more prevalent in scientific research due to the advancement and innovation of new computing techniques and algorithms. To conveniently present data, graphical methods are used with various visualization tools. Graph theory is one of the primary approaches to describing objects and situations. Graphical representations of models are essential for rapidly comprehending, absorbing, and conveying critical information. In mathematical terms, a graph consists of a finite set of vertex points that can be linked by edge lines. Today, no scientific field can function effectively without using graph models. The field of information technology relies on multifaceted object interactions, making broadcasting information through this kind of method vital [1]. The article will delve into graph structures and situational analysis, which present the foundation for developing models of logical and conceptual representation of proposals.

Understanding textual constructions in the Russian language is an intricate and multifaceted procedure. Most current intelligent assistants rely on conditional structures and question-answer templates. These rudimentary algorithms underpin rapidly developing systems that now boast vast and regularly updated databases [2]. Artificial limitations in communication with AI reduce interest in interacting with such systems, resulting in the exclusive use of standard phrases. The processing of textual information is currently hindered by information noise, which refers to the excess of trivial information cluttering internet resources. Natural language processing models have advanced significantly and are applied in various fields. There are numerous tasks that analyzers can solve. To tackle the issue of semantic allocation of structures for future use, we employ conditional models, despite their low accuracy. It's worth noting that these methods are limited by specific language rules and conditions. To enhance the identification of further parameters and characteristics for each analyzed structure, it is advisable to employ graphical representations. Graph-based natural language processing models offer an intriguing solution for selecting additional criteria and promptly detecting relevant information.

In connection with the above areas, a number of studies of models and techniques will be conducted, according to which it will be possible to organize an automated analyzer of Russian language sentences based on graph structures.

2 Materials and Methods

2.1 Graph Core and Function Levels

Graph structures are commonly used in machine learning models, representing a specific approach to function development.

The graph kernel is a concept that describes graph behavior in relation to homomorphism. Homomorphism maps data between two graphs, while each graph's structure remains unchanged. It is possible to establish the similarity between graphs, as depicted in Fig. 1. A graph is considered a kernel when every homomorphism is an isomorphism. This allows the performance of selected operations within the graph, regardless of the space surrounding it. For the design of the proposed system, this method is essential to emphasizing the most significant aspects of language. It is crucial to ascertain the potency of every term within its context to produce a relevant and significant solution.

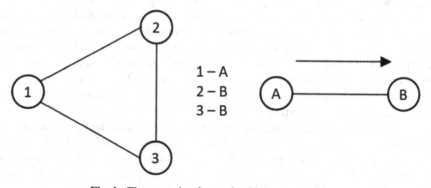

Fig. 1. The example of a graph with homomorphism.

Homomorphism enables simplification of the structure and reduction of adjacent vertices. They constitute a specific aspect of graph representation that emphasizes the most significant features of the subject domain [3]. Subsequently, we will examine graph isomorphism, which is one of the homomorphism instances.

An isomorphism between two graphs represents a one-to-one correspondence between their sets of vertices, such that the adjacency of any two vertices in graph A is the same as that of their corresponding vertices in graph B. An illustration of this concept is provided in Fig. 2. The graphs under consideration may be undirected and lack weights for both their vertices and edges. The isomorphism technique is valuable in processing semantic constructions, as it enables the identification of analogous sentences. It also supports the selection of appropriate data for training neural networks.

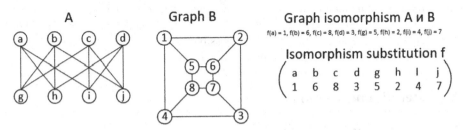

Fig. 2. Isomorphism on the example of a graph.

Isomorphism expresses the similarity or coincidence of two graphs in both directions. The use of homomorphism in algorithms for processing semantic constructions enables detailed processing and identification of the most significant sentences. Applying a preorder and identifying dependencies among vertices of graphs enables the creation of a system that uses predetermined knowledge bases to process incoming requests and provide an answer promptly [4–6]. However, defining the processing area is crucial in tackling the problem of isomorphism, as the algorithm's search for such isomorphism may take too long and result in no response from the system.

The proposed system employs statistical aggregation at the node level, providing a straightforward method of defining a function at the graph level. The proposed system employs statistical aggregation at the node level, providing a straightforward method of defining a function at the graph level. This compiled information is then utilized to represent the graph at its highest level. Consequently, statistics can be conveniently consolidated at the node level. Frequency histograms and other summary statistics can then be calculated based on word usage, node centrality, and clustering coefficients in the graph. One drawback of this method is its reliance on local node data, potentially omitting critical global properties of the graph.

An iterative neighbourhood aggregation strategy can enhance the fundamental technique with a node set. These methodologies seek to extract node-level functions with more informative content beyond local data. Feature selection functions are combined in the graph-level representation. The aggregation process concludes upon achieving a perfect model, with different parts of speech assigned to each property type.

The Weisfeiler-Lehman algorithm and kernel are two of the most significant and widely recognized aggregation strategies. Typically, this label corresponds to a degree in most graphs, [7–9]. The algorithm's main concept involves the assignment of an initial label to each node. Subsequently, every node is assigned a new label iteratively. This new label is generated by converting the multisets of current neighbouring node labels into a fixed-length bit string.

$$l^{(i)}(v) = HASH\left(\left\{\left\{l^{(i-1)}(v) \forall v \in N(v)\right\}\right\}\right). \tag{1}$$

When double curly brackets indicate a multiset, the HASH function assigns a unique new label to each unique multiset. The hash function transforms input data, of any length, into an output bit string of a specific length. After several rounds of highlighting properties, a label emerges for each node which summarises the structure of its surrounding area. Consequently, summary statistics can be calculated on these labels as a functional representation of the graph. The Weisfeiler-Lehman kernel is computed by quantifying the dissimilarity between the outcome sets of labels for two graphs. The Weisfeiler-Lehmann kernel is significant from a theoretical viewpoint [10]. To estimate graph isomorphism, one can examine whether two graphs share the same property set after several rounds of the mechanism. This technique resolves some isomorphism difficulties when processing sentence constructions on graphs.

2.2 Graphlet-Based Methods

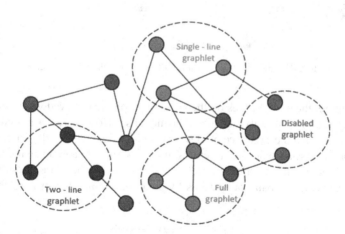

Fig. 3. Types of assigned graphlets.

When analysing functions at the node level, our approach involves counting the frequency of small subgraph structures present in the constructed simulations of each sentence. These structures are known as graphlets and are isomorphism classes of induced subgraphs in a graph. The graphite core formalizes this method by listing all possible graph structures of a specific size and determining the number of occurrences in a complete graph. Figure 3 illustrates a range of graphlets for comprehending the allotted structures.

This technique encounters issues with intricate graphlet counting, necessitating computational power even after pre-processing each situation with a set of approximations.

2.3 Metric of Similarity of Neighbors

The examined approaches for extracting features or statistics of individual nodes and entire graphs have practical applications in the envisaged model for visually classifying vital parameters. Nonetheless, they possess limitations since they fail to quantify internode relationships [11–14]. This indicates that the aforementioned methods are insufficient for forecasting connections necessary for graphs that can be dynamically scaled and for modelling response sentences textually. The objective is to predict the existence of an edge between two nodes, as depicted in Fig. 4.

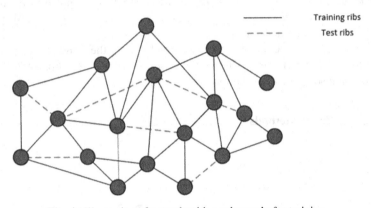

Fig. 4. Illustration of a graph with a subsample for training.

In the figure, the dotted edges on the training graph are removed when training the model or calculating similarity statistics. The model's performance is assessed based on its aptitude to predict the existence of these evaluation edges. As words can be placed at any position in Russian sentence structures, it is imperative to anticipate the maximum feasible manoeuvres for positioning each node. One way to measure the similarity of neighbourhoods among pairs of nodes is by counting the number of neighbours they share.

$$S[u, v] = |N(u) \cap N(v)|. \tag{2}$$

where $S[u,v]$ represents the value defining the relationship between nodes u and v. To avoid overburdening the model with machine learning techniques, this approach enables the rapid and efficient prediction of most connections in a large data stream. The fundamental principle of the similarity statistics of neighbours $S[u,v]$ is to assume that the probability of an edge (u,v) is proportional to $S[u,v]$ [8].

$$P(S[u, v] = 1)\alpha S[u, v]. \tag{3}$$

Thus, to address the issue of prediction of relationships using a measure of similarity of neighbours, you need to set a threshold to determine when the existence of an edge should be predicted.

2.4 Model-Parametric Representation

Graph theory is the most suitable tool for describing and investigating the structure of a space that contains models, parameters, and their relationships. One can refer to the subspace of parameters associated with models or the subspace of models associated with parameters. By using weighted digraphs, we can establish the structure of this space and apply specific actions based on the obtained results. It is feasible to identify additional subspaces within a specified space based on varied traits [15–17].

Based on the image of the model-parametric structure, we will create a model for situational analysis of the information representation using neighbourhood selection. The first-order spaces described above are the neighbourhood of parameters directly linked to the model. It is imperative to assign this neighbourhood as a distinct structure. All components of the model's vicinity, as asserted, are factors and are spaced apart by one unit.

A model-neighbourhood comprises all the model-parametric space elements that are connected via paths. The model neighbourhood boundary consists of space elements that are linked to the model in one direction. Figure 5 illustrates an example of such a space.

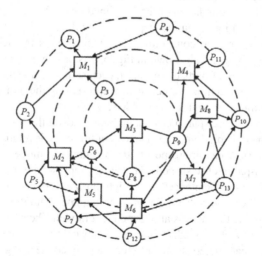

Fig. 5. Model-parametric space relative to the M3 model.

By analogy with model districts, parameter regions exist. Utilising regions is crucial in representing structures for subsequent analysis with graphs [18]. We will refer to those elements belonging to each graph region's neighbourhood as the intersection of the space under review. Figure 6 demonstrates an example of this intersection.

From the example, it is evident that the intersection of neighbourhoods does not result in a neighbourhood. To identify patterns and conduct automated analysis of proposals,

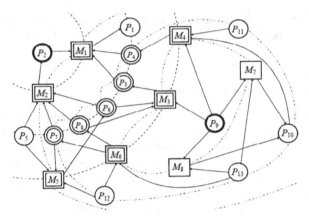

Fig. 6. Results of the operation of intersection of model neighbourhoods.

it is essential to recognise predetermined structures that are logical. Such structures are called graphlets, representing certain situations in the legal space of interaction between models and parameters of the designed environment. The graphlets chosen are demonstrated in Fig. 3.

This method struggles with intricate graphlet counting, requiring substantial computing power, and even pre-processing each situation with a set of approximations. Consequently, we present the foundation of the examined texts. We have produced a model of the abstract representation of our concept. Utilising the graph depicted in Fig. 6, we can establish the structure and connections of sentence parts in each sentence being appraised. According to the graph shown in Fig. 6, it is possible to determine the structure and interrelationships of parts of speech in each sentence under consideration. By comparing the parts of speech in a sentence and representing them in a graph, semantic elements can be combined to highlight graphlets.

When constructing model-parametric neighbourhoods at the logical level, a particular model or parameter is predominant. The rest of the space is relative to this element, and its ordering is carried out with respect to the parameter or model being studied [19–21]. Thus, the model-parametric space is the primary representation when constructing models for Russian language sentences. The neighbourhoods of each subject could serve as potential links between objects and parameters in a given situation.

Each sentence in this text acts as an independent model. The presented architecture suggests that some parameters of adjacent models may coincide or have a similar meaning. As a result, these parameters connect to form a fully connected graph. To highlight the current situations and identify the direction of the sentence and its semantic load, we use constructions predefined in meaning, called graphlets. This perspective can be enhanced and expanded in conjunction with the handling of new resources. The primary trait is the brevity and graphical depiction of the desired models' various versions. Additionally, there is a chance to conceive the required circumstance in any field of study, courtesy of a user-friendly design framework. The exploration of developing a system for presenting textual information using model-parametric models enables the structuring of text through graphical representation. The architectures designed within the model-parametric space are applicable to software that analyses Russian sentences.

3   Results and Discussion

Based on the previously discussed methods, Fig. 7 demonstrates a created syntactic graph model. The syntactic graph abstractly represents the distribution of parts of speech within a sentence for logical purposes. The interpretation of each part of speech depends on its frequency of use in the text. The frequency of individual word usage is calculated during the training iterations. Afterward, the words are classified by their parts of speech and their frequency characteristics and graphlets are determined. After determining the semantic purpose of the words, the identifier of the logical role in the sentence is assigned.

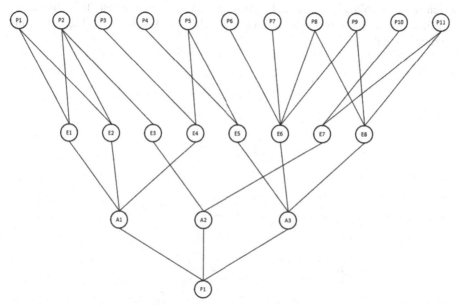

Fig. 7. Syntactic graph: In Fig. 7, P1 – noun; P2 – pronoun; P3 – verb; P4 – participle; P5 – adverb; P6 – adverb; P7 – preposition; P8 – particle; P9 – conjunction; P10 – adjective; P11 – numeral; E1 – subject of action; E2 – object of action; E3 – component of action; E4 – action; E5 – interaction; E6 – relation; E7 – set of properties; E8 – relation; A1 – main idea; A2 – description; A3 – logical structure. F1 is the resulting expression. The resulting expression will be an entire sentence. It is divided into valid constructions by meaning: the main idea (A1), description (A2), logical structures (A3). Each construction expresses a specific element of the graph (E1 – E8). The element is a representation of a part of speech (P1 – P11).

It is essential to handle each active construction in distinct ways and algorithms of varying complexities, depending on the text's saturation. The existing design indicates an element or elements of the conceptual structure displayed in Fig. 8.

The element portrays specific parts of speech in a sentence that are categorized for further scrutiny. The demonstrated interaction enables one to deconstruct propositions into simpler constructions for easy handling. Additionally, this construction enables the generation of straightforward spoken expressions through a trained model. This has led to the suggestion of combining the produced statements into complete sentences based on the structure of the syntactic graph.

In order to verify the obtained graphs, the existing models of natural language processing were redesigned. The introduction of preliminary results showed that when using a syntactic graph, additional connections appear based on the graph of the conceptual structure. These improvements allow you to get a better result when processing small texts. To improve the accuracy of algorithms for analysing large texts, other, more significant changes should be made. Comparison with existing models of natural language processing was carried out using software implementation of processing algorithms and implementation of a modified model with additional criteria. When compiling the modified model, graph isomorphism, graphlets and the metric of similarity of neighbours, which were described in the second section, were taken into account. This allowed us to achieve better results when used on a text smaller than 10000 words. The test results are shown in Table 1.

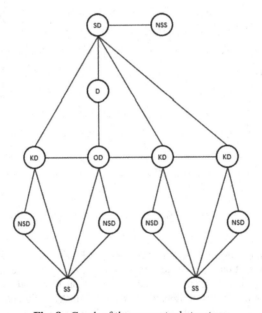

Fig. 8. Graph of the conceptual structure:

In Fig. 8, NSS is the set of properties of the subject; SD is the Subject of the action; D is the Action; OD is the Object of the action; KD is the Component of the action; NSD is the Set of properties of the action; SS is the ratio of properties.

The results shown in Table 1 demonstrate the outcomes of testing various natural language processors on Russian text. In order to compare results across other languages, it would be necessary to adjust the model and metrics accordingly; therefore, this study solely considers and tests architecture in Russian. The table data demonstrates that the

Table 1. Implementation of the results obtained.

The volume of the text Model	5000–10000 words, % accuracy	15000–50000 words, % accuracy	60000–150000 words, % accuracy
RoBERT	54.2	59.1	72.4
DistilBERT	56.3	57.2	73.9
XLM - RoBERTa	63.3	73.9	85.8
Modified model	69.6	54.4	57.3

modified model based on the syntactic graph delivers the best results in processing short text. With medium and large text, the accuracy is significantly reduced, for various reasons. There is an oversaturation of certain words and speech constructions, and words that are rarely used are lost against their background. Large speech models use many factors and conditions to avoid such situations.

4   Conclusions

As a result, it can be concluded that the use of various methods of simplifying complex structures allows you to unload the analysed model to speed up the processing time of information. A unique graph model of the distribution of significant sentence structures for subsequent analysis and textual modelling, called a syntactic graph, is proposed. The methods discussed in the article, at the initial stage of processing, will allow sorting words and phrases in text constructions according to a variety of criteria. The syntactic graph is modelled in accordance with the speech turns of the Russian language and allows you to uniquely determine the role of a specific part of speech in the context. Natural language processing models are currently very diverse and are used in various fields. There are many types of tasks that are solved by different types of analysers. To solve the problems of semantic allocation of structures for further work with them, conditional models are used, which have low accuracy. Also, such methods of obtaining a result are very limited by the rules that are capable of searching for speech constructions of a certain language under certain conditions. To increase the number of additional parameters and characteristics of each analysed structure, graphical representations of information can be used. Graph-based natural language processing models can be an interesting solution when selecting additional criteria and quickly detecting the necessary information.

According to the results of the research, it was possible to identify the optimum conceptual structure, upon which an algorithm for processing semantic constructions without utilising condition models was developed. The result of checking the designed model shows that the utilisation of a syntactic graph in the distribution of parts of speech at the initial stage of analysis enables you to identify logical constructions. Based on the obtained patterns, additional criteria are assigned to each word in order to enhance the accuracy of further text determination. However, as the number of processed words increases, the model's accuracy declines as words that are infrequent but still relevant to a particular text are assigned less weight. According to the results of the research, it was revealed that the developed model can only work with texts in Russian because the logical constructions and features of constructing sentences in other languages will be very different. The rules that are included in the algorithm must be adjusted according to the syntax of the specific language under study. Further improvements of this algorithm allow you to determine the necessary structure for a different language and type of narrative. The designed analyser will become part of the system of semantic information processing and modelling of text structures. The architecture of further processing may differ from existing models to achieve the best results. Thus, using many simple methods of data ordering and analysis, you can get a fast-automated tool for working with text.

References

1. Anferov, M.A.: Genetic clustering algorithm. Russ. Technol. J. **6**(7), 134–150 (2019)
2. Sorokin, A.B., Lobanov, D.A.: Conceptual design of intelligent systems. Inf. Technol. **1**(24), 3–10 (2018)
3. Khurana, D., Koli, A., Khatter, K.: Natural language processing: state of the art, current trends and challenges. Multimed. Tools Appl. **82**, 3713–3744 (2023)
4. Sorokin A.B., Smolyaninova V.A.: Generalized integrated characteristic base of modular number system **9**(23), 634–641 (2017)
5. Krasnikov, K.E.: Mathematical modeling of some social processes using game-theoretic approaches and making managerial decisions based on them. Russ. Technol. J. **9**(5), 67–83 (2021)
6. Zhang, X., Zhao, H., Chen, D.-Y.: Semantic mapping methods between expert view and ontology view. J. Softw. **31**(9), 2855–2882 (2020)
7. Sydorenko, V., Kravchenko, S., Rychok, Y., Zeman, K.: Method of classification of tonal estimations time series in problems of intellectual analysis of text content. Transp. Res. Procedia **44**, 102–109 (2020)
8. Sorokin, A.B., Zheleznyak, L.M., Suprunenko, D.V., Kholmogorov, V.V.: Designing modules of system dynamics in decision support systems. Russ. Technol. J. **10**(4), 18–26 (2022)
9. Anferov, M.A.: Algorithm for searching subcritical paths on network graphs. Russ. Technol. J. **11**(1), 60–69 (2023)
10. Tomashevskaya, V.S., Yakovlev, D.A.: Methods of processing unstructured data. Russ. Technol. J. **9**(1), 7–17 (2021)
11. Tatur, M.M., Lukashevich, M.M., Pertsev, D.Y., Iskra, N.A.: Data mining and cloud computing. Doklady BGUIR **6**(124), 62–71 (2019)
12. Sobolevsky, S., Belyi, A.: Graph neural network inspired algorithm for unsupervised network community detection. Appl. Netw. Sci. **7**(63) (2022)
13. Kochkarov, R.A.: Research of NP-complete problems in the class of prefractal graphs. Mathematics **9**(21), 2764 (2023)
14. Liu, B., et al: Graph neural networks in natural language processing: a survey. Now Found. Trends (2023)
15. le Gorrec, L., Knight, P.A., Caen, A.: Learning network embeddings using small graphlets. Soc. Netw. Anal. Min. **12**, 12–20 (2022)
16. Wang, H., Li, J., Wu, H., Hovy, E., Sun, Y.: Pre-trained language models and their applications. Engineering **25**, 51–65 (2023)
17. Zhou, J., et al: Graph neural networks: a review of methods and applications. AI Open **1**, 57–81 (2020)

18. Belov, S.D., Matrelova, D.P., Matrelov, P.V., Korenkov, V.V.: Review of methods of automatic text processing in natural language. Syst. Anal. Sci. Educ. **3**, 8–22 (2020)
19. Sadovskaya, L.L., Guskov, A.E., Kosyakov, D.V., Mukhamediev, R.I.: Text processing in natural language: a review of publications. Artif. Intell. Decis.-Mak. **3**, 66–86 (2021)

Information and Computing Technologies in Automation and Control Science

Classic and Modern Methods of Automatic Parking Control of Self-driving Cars

Ilya D. Tyulenev[1] and Nikolay B. Filimonov[1,2(⊠)]

[1] Lomonosov Moscow State University, Leninskie Gory, 1, 119991 Moscow, Russia
tiulenev.id19@physics.msu.ru, nbfilimonov@mail.ru
[2] Bauman University, 2nd Baumanskaya Street, 5/1, 105005 Moscow, Russia

Abstract. The problem of automatic parking control of an unmanned car is considered. The formulation and formalization of the problem of the car parking control taking into account the restrictions that ensure the safety of the parking maneuver are given. Classical and modern methods of the automatic parking control of unmanned cars are analyzed. Based on the Dubins and Reeds-Shepp motion models, optimal algorithms developed for the car parking control are synthesized. A fast-growing random tree algorithm RRT is used to construct a path between two points. Based on the method of machine learning involving reinforcement, a car parking control algorithm is synthesized. The algorithm convergence is investigated, and the optimal values of the training parameters are determined. The results of the computer testing of synthesized parking algorithms implemented in Python using mathematical libraries Matplotlib and NumPy are presented.

Keywords: Self-Driving Car · Automatic Parking Control · High-Speed Parking Trajectories · Dubins and Reeds-Shepp Motion Models · Reinforcement Machine Learning · Q-Learning

1 Introduction

Nowadays the developments of a self-driving car (SDC) equipped with an automatic pilot, that is with an automatic system ensuring the traffic control of the car without the driver's participation [1], have gained the great popularity. In this case, a full autonomy of SDC is achieved by automating the control of all its movement conditions and maneuvers, including perhaps the most prevalent maneuver such as parking [2]. The problem of the parking automation has assumed the peculiar urgency because it allows not only facilitating the process of safety parking, but also increasing the density of packed cars of 62–87% [3].

There are two main directions of modern development of SDC. The first direction, which is followed by the majority of automobile companies (Tesla, BMW, Audi, Mercedes), assumes gradual achievement of increasingly higher levels of automobile driving automation: starting from auxiliary passive systems and ending with a fully autonomous car. The Society of Automotive Engineers "SAE international" has put forward the concept of six levels of autonomy of unmanned cars. Navigant Research analysts predict

V. Jordan et al. (Eds.): HPCST 2023, CCIS 1986, pp. 197–209, 2024.
https://doi.org/10.1007/978-3-031-51057-1_15

that 75% of the cars sold in 2035 will have some sort of an autonomous capability. The second direction, put forward by Google, involves the Street View-based development of fully automated cars without the ability to control human driving. Ultimately, both approaches envision achieving a full level of automation where no human will be required to drive the car [4].

The first systems of automation parking were created in the mid-2000. In spite of their impetuous development, the classical parking automation control methods are still very popular. They are based on the principles of the geometric construction of support trajectories of the moving objects (cars, mobile robots, ships, drones etc.) on the plane. These methods correspond to the minimum time of the object movement from the initial configuration to the specified final configuration based on the motion model of the Dubins machine and its modifications [5]. However, in recent years, modern, "trainable" methods of the SDC automatic parking control, based on using artificial intelligence methods and technologies: genetic algorithms, fuzzy logic, neural networks, Internet of things technologies, etc. (see, e.g. [15−23]) have been used rather than classical, "untrainable" methods. This paper develops the results of [24, 25]. The issues of computer synthesis and research of the algorithms intended for the optimal speed control of parking SDC based on the classical control method using the Dubinse and Reeds-Shepp motion models and the modern control method using reinforced machine learning, providing automatic "smart" car parking are analyzed.

2 Peculiarities of the Parking Control Problem

2.1 Equations of SDC Parking Dynamics

Let us introduce the following notations. $W = (x, y)$ is a coordinate of the center of the front wheelbase of the car in the coordinate system XOY; θ is an angle between the axis OX and the straight, passing over the centers of back and front wheelbases of the car; ϕ is the rotation angle of front wheels, and v is the car speed (see Fig. 1).

Let us suppose that during the parking process, the car moves slowly, without high motor rotations and sliding wheels. Then the mathematical model of the SDC parking dynamics may be represented as the following system of the differential equations:

$$\dot{x} = v\cos(\theta + \phi), \tag{1}$$

$$\dot{y} = v\sin(\theta + \phi), \tag{2}$$

$$\dot{\theta} = v\frac{\mathrm{tg}\phi}{l}. \tag{3}$$

The car state is characterized by a three-dimensional vector in the form of:

$$\mathbf{s} = (x, y, \theta)^{\mathrm{T}}, \tag{4}$$

and its control is provided by a two-dimensional control vector in the form of:

$$\mathbf{u} = (v, \phi)^{\mathrm{T}}. \tag{5}$$

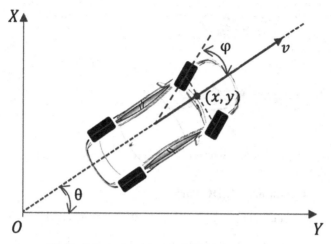

Fig. 1 Calculation scheme of the SDC movement.

2.2 Mechanical Limitations of SDC Parking

The smoothness of the car-parking trajectory, that is, the limitation of its speed:

$$\dot{k} = \frac{\omega(t)\cos\phi(t)}{l},$$

is provided by the specified limitations imposed on the wheel rotation angle, and also on the linear and angular speed of the car:

$$\varphi(t) \leq \varphi_{max}, v(t) \leq v_{max}, \omega(t) = \dot{\varphi}(t) \leq \omega_{max}. \tag{6}$$

2.3 Space Limitations of SDC Parking

Secure car parking is provided by space limitations, permitting to avoid its collision with the enclosing objects that are present in the parking zone. In order to avoid the mentioned collision, let us introduce the circles of "safety", covering the car and the obstacles in the parking zone (Fig. 2).

In doing so, let us suppose that the car is covered by N 'safety' circles of R_{C_i}, a radius having centers $C_i, i = 1 : N$, and the obstacles are covered by M 'safety' circles of the R_{O_j} radius having centers $O_j, j = 1 : M$. Evidently, avoiding the car collision with the obstacles allows the fulfilment of the following conditions:

$$\left|C_i O_j\right| > R_{C_i} + R_{O_j}, i = 1 : N, j = 1 : M, \tag{7}$$

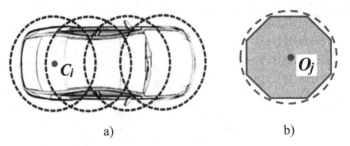

a)   b)

Fig. 2 "Safety" circles i of SDC (a) and an j obstacle (b).

2.4 Boundary Conditions of SDC Parking

Let us suppose that at the start of parking $t = t_0$, the car is in an arbitrary state:

$$\mathbf{s}(t_0) = (x(t_0) = x_0, y(t_0) = y_0, \theta(t_0) = \theta_0), \tag{8}$$

$$v(t_0) = 0, \phi(t_0) = \phi_0, \tag{9}$$

and at the end of parking $t = t_f$, the car is in a target state:

$$\mathbf{s}(t_f) = (x(t_f) = x_f, y(t_f) = y_f, \theta(t_f) = \theta_f), \tag{10}$$

$$v(t_f) = 0, \phi(t_f) = 0. \tag{11}$$

2.5 Setting the SDC Parking Control Problem

The car of a specified dynamics model (1)–(3) requires synthesizing a control algorithm ensuring its automatic secure parking, that is, a transfer from the arbitrary initial state (8), (9) to the goal final (10), (11) state for the minimum time:

$$t_f - t_0 \rightarrow \min$$

with regard to mechanical (6) and space limitations (7).

In the recent automatic control theory, the mentioned problem is related to a class of limit fast acting problems [5].

Among the control methods, during the fast acting car maneuvers, the methods based on Dubins and Reeds-Shepp models movement are very popular (see e.g. [26, 27]). These maneuvers of the car include the parking in the limiting space having static obstacles in conditions of limit controllability by the angular speed.

3 Parking Based on Dubins and Reeds-Shepp Motion Models

The Dubins motion model (L.E. Dubins) has the following form [26]:

$$\begin{cases} \dot{x}(t) = v(t)\cos"(t), \\ \dot{y}(t) = v(t)\sin"(t), \\ \dot{\theta}(t) = \omega(t) \le \frac{1}{R}. \end{cases} v(t) = 1.$$

And according to the Pontryagin's maximum principle, it corresponds to the limit in the fast acting system's transfer (13) from one condition to another by piecewise constant control.

According to Dubins, the following six combinations of control are optimal: *RSR*, *LSL*, *RSL*, *LSR*, *RLR* and *LRL*. Here the following notations are specified: "*R*" is to turn right by the smallest radius, "*L*" is to turn left by the smallest radius, "*S*" to is move straight. An example of the SDC trajectory simulation when parking along the trajectory of Dubins is represented in Fig. 3a. The Reeds-Shepp motion model (J.A. Reeds & L.A. Shepp) is the modification of the Dubins model allowing the car both forward and backward motions [27]. The above motion model is responsible for parking the car more fully. An example of the SDC motion trajectory simulation, when parking the car along the trajectory of Reeds-Shepp, is represented in Fig. 3b.

The computer approbation results of simulation algorithms of SDC parking (see Fig. 3) show that the utilization of Reeds-Shepp motion model allows fulfilling the parking within the smaller free space using the Dubins motion model.

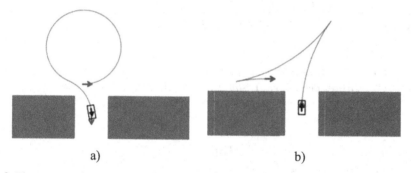

a)   b)

Fig. 3 The comparative analysis of the SDC parking control method based on Dubins (a) and Reeds-Shepp (b) motion models.

The main advantages of using Dubins and Reeds-Shepp motion models are their limit fast acting, the simplicity of calculating support trajectories, the possibility of choosing other (non-optimal) trajectories in the presence of obstacles and their applicability for any moving objects having a limit turning radius.

The trajectories of the Dubins motion model are noted to have a non-symmetry because of its unidirectional motion (only forward). In this connection, the Reeds-Shepp motion model, being bidirectional (allowing a backward motion) and having a symmetry, is of interest in the case of applied control problems by maneuvering moving objects including SDC parking.

The Reeds-Sheep model may be used as a base model in constructing the optimal support car trajectories of the movement containing many switching points and a sequence avoiding obstacles. In this case, the sampling methods of traffic planning [28], where the algorithms of a rapidly exploring random tree (RRT) [29] are the most effective, have the greatest popularity. By virtue of randomization, the independence from the geometric representation and the simulated environment dimension is the important merit for the mentioned algorithms.

The Reeds-Shepp motion model is used in constructing a rapidly random tree RRT. According to the given RRT method, the search tree $T(V, E)$, whose roat V_s is the initial conditions of the car, is growing by means of adding new tops chosen accidentally from the free search area. In this case, the new top V_{new} is added to the tree, provided that the car transfer in this condition satisfies the introduced space and mechanical limits corresponding to the tree point V_{obs}. When the car has the radius limitation of the trajectory curvature, the Reeds-Shepp motion trajectories can be selected as the shortest ways between the nearest tree point V_{near} and the random point V_{rand}. The algorithm is finished by coincidence with the specified exactness of the last top added in the tree and the top of the target state V_f. The pseudocode of the algorithm used for constructing the trajectories of the SDC movement, when parking by means of the Reeds-Shepp motion model and quickly exploring the tree, has the following form:

```
 1: Given: Vₛ, V_f, eps.threshold
 2: Until then Distance(V_new, V_f) > r to fulfill
 3: T ← V_new
 4: T₀ generate V_rand
 5: T₁ find V_near ∈ T: min|V_near, V_rand|
 6: If Distance(V_near, V_abs) > threshold then
 7: To construct E(V_near, V_rand) ∈ T (Reeds-Shepp way)
 8: Otherwise
 9: To generate  V_rand
10: The end of condition
11: The end of cycle
12: To return
```

The main advantages of the algorithms based on quickly exploring the tree are their simplicity (the simplicity of realization and availability of simple calculations), fast convergence and extension to unexplored areas of the free space, and an asymptotic guarantee of finding a solution. The constructed tree permits to construct the support trajectories of the car parking when maneuvering in the limited space having obstacles. When moving the car along the constructed support trajectory of the car parking, its transfer from the arbitrary initial state to the target final state in the presence of the mentioned mechanical and space restrictions is provided.

The computer approbation of the above algorithm realized in the Rython language using the mathematical libraries NumPy and Matplotlibis represented in Fig. 4 (the obstacles are highlighted in red).

The computer experiments conducted on the SDC dynamics model have confirmed the effectiveness of the suggested algorithmic solutions of the automatic controlling of its parking.

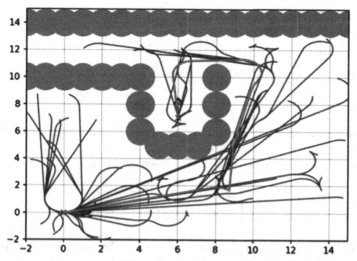

Fig. 4 Computer approbation of the SDC parking control algorithm based on the RRT method using the Reeds-Shepp motion model.

4 Parking Based on Reinforcement Learning

The learning that involves reinforcement (Reinforcement Learning, RL) is the area of the machine learning, which is the perspective direction for the development of the effective, universal and realizable control algorithms by the complex moving objects providing improved skills and adaptation of the object behavior to conditions of its functioning in an uncertain work environment [30, 31].

The work [32] considers the application of the machine learning without a teacher based on neuron networks aimed at the development double controller providing the transverse and longitudinal maneuvering of the car intended for automatic parking in a confined space. The parking algorithm based on recursion neuron networks and the approaches of machine learning is represented in [33]. The methods of the machine reinforcement learning are of greater interest. Their use allows bringing SDC maneuvering as close as possible to the generally professional parking control realized by the driver [34, 35].

In the conception of machine reinforcement learning, SDC is considered as an intellectual agent interacting with environment. The aim of the agent is to maximize some criterion named as "a reward" and received in the learning process as a compensation for learning in the form of a feedback signal obtained from the environment intended for the realization of fast, exact and secure SDC parking. The strategy of the agent's action is formed automatically based on the obtained information about the results of the actions.

One of the most effective learning reinforcement algorithms is an algorithm of Q-learning suggested by Watkins (C.J.C.H. Watkins) in 1989. This algorithm uses a greedy strategy when choosing the agent's actions and allows defining the optimal choice of the actions for any state [36].

The agent state at the step t is described by the state vector:

$$\mathbf{s}_t = (x(t), y(t), \theta(t))^{\mathrm{T}}.$$

Performing the action $\mathbf{a}_t$, SDC goes from a state $\mathbf{s}_t$ to a state $\mathbf{s}_{t+1}$:

$$\mathbf{a}_t(v, \phi) : \mathbf{s}_t \rightarrow \mathbf{s}_{t+1}.$$

Let us introduce the utility function, that is a matrix $Q(\mathbf{s}, \mathbf{a})$, which stores the quality score connected with every "state-action".

The action $\mathbf{a}_t$ is selected according to the "ε-greedy" principle:

$$\mathbf{a}_t = \begin{cases} \text{random}, & 0 \leq n < \varepsilon, \\ \underset{\mathbf{a}}{\text{argmax}}\, Q(\mathbf{s}, \mathbf{a}), & \varepsilon \leq n < 1. \end{cases} \tag{12}$$

A set of actions $[\mathbf{a}_{t_0} \ldots \mathbf{a}_{t_f}]$ provides the motion of the vector $\mathbf{s}_t$, whose projection onto a plane XOY is the movement trajectory of the car $k(t)$.

Let us introduce the function $R(\mathbf{a}_t)$, being the momentary reward, returned by the environment as the estimation of the last action $\mathbf{a}_t$:

$$R(\mathbf{a}_t) = \begin{cases} 1000, & \mathbf{s}_{t+1} = \mathbf{s}_{\text{final}}, \\ -20, & \mathbf{s}_{t+1} \neq \mathbf{s}_{\text{final}}, \\ -300, & C_i \cap O_j. \end{cases}$$

The first expression in this system defines the big reward used for reaching the final SDC position in the parking place; the second expression defines penalty for every maneuver not carrying to the aim; and the third one defines the big penalty for collision with the obstacle. The machine learning scheme involving reinforcement is represented in Fig. 5.

The reward matrix Q is filled iteratively, maximizing the expected award at the following step $\mathbf{s}_{t+1}$ from the current state $\mathbf{s}_t$ according to the Bellman equation:

$$Q(\mathbf{s}, \mathbf{a}) = (1 - \alpha)Q(\mathbf{s}, \mathbf{a}) + \alpha(R(a_t) + \gamma \max_{\mathbf{a}_{t+1}} Q(\mathbf{s}_{t+1}, \mathbf{a}_{t+1}))$$

where α is the speed of learning, preventing a fast change Q in one renewal, γ is the factor of trust in future rewards. Let us note that in the case of large values α, the algorithm may terminate too quickly having reached a local minimum, and as for the small values γ, the algorithm becomes a "greedy" one.

Filling in the matrix Q evidently does not guarantee the implementation of the Bellman's optimality principle. In fact, according to the relation (12), the agent's choice of the action $\mathbf{a}_t$ allows at some frequency committing random actions aimed to investigate the learning environment more effective in order to find the new optimal parking space trajectories.

Let I be the number of parking episodes in the learning process and the number of actions t be performed by an agent when $0 < t \leq 30$. Let us introduce the parameter

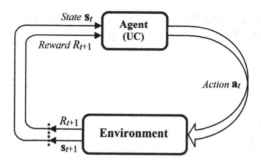

Fig. 5 Machine learning scheme involving reinforcement.

setting the quality assessment of the completed learning which is equal to the relation of the number of successfully completed maneuvers to the general number of maneuvers:

$$\eta = \frac{I_{\text{success}}}{I_{\text{all}}}. \tag{13}$$

Table 1 presents the parameters of machine learning and their optimal notes determined in the course of the study. The evaluation results of the parking by the trained model intended for different values of parameters α and γ are represented in Table 2.

Table 1. Parameters of machine learning.

Parameter	Name	Tolerance interval	Value
α	Learning rate	[0, 1]	0.3
γ	Discount factor		0.75
ε	Exploration factor		0.5

Table 2. Success rate for various values of parameters.

α	0	0.15	0.30	0.45	0.60	0.75	0.90	1.00
η, %	–	92	99	95	91	87	82	77
γ	0	0.25	0.5	0.75	1.00	–	–	–
η, %	–	92	95	99	97	–	–	–

To assess the algorithm quality, its convergence was investigated using the function:

$$\delta_k = \begin{cases} C = \text{const}, \ k = 0, \\ \delta_{k-1}, \ k > 0, \ \not\exists s_{t+1}, \\ (1 - \xi)\delta_{k-1} + \xi |\, Q_k(\mathbf{s}, \mathbf{a}) - Q_{k-1}(\mathbf{s}, \mathbf{a}) \,|, \ k > 0, \ \exists s_{t+1}. \end{cases}$$

The graph in Fig. 6 demonstrates the consequent values of δ_k having an increasing number of iterations; the following values of the function Q practically stop changing compared to the previous one. Hence, the learning process can be stopped, and we can assume that the agent (SDC) has learnt how to park.

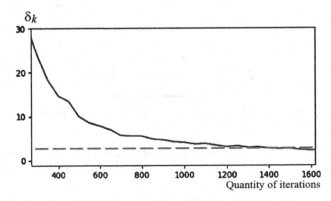

Fig. 6 Convergence of the algorithm.

Table 3 shows the dependence of the percentage of successful maneuvers on the number of learning iterations and provides the average maneuver execution time that SDC wastes when it parks from the arbitrary initial position.

An example of the computer simulation of automatic parallel SDC parking by the machine learning method involving reinforcement is represented in Fig. 7.

Table 3. Successful maneuvers for various number of learning iterations.

$I \cdot 10^6$	η, %	t
0.4	5	5
0.85	29	11
1.2	82	12
1.6	97	9
2.0	99	9
2.4	100	9

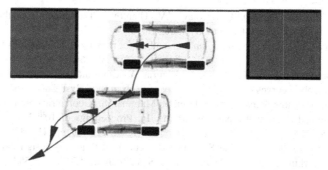

Fig. 7 Parallel SDC parking by the machine learning method involving reinforcement.

5 Conclusion

The control problem of SDC automatic parking, taking into account the restrictions ensuring the parking maneuver safety, is considered. Control algorithms realizing both classical and modern methods of automatic car parking are proposed.

The classical control method of SDC parking is based on the utilization of widespread Dubins and Reeds-Shepp traffic models that provide optimal car parking in terms of the speed. Control algorithms of car parking realizing the classical control method of car parking are suggested.

The modern control method of SDC parking is based on using intelligent technologies. The control algorithm of car parking realizing the modern method of parking based on machine learning involving reinforcement is proposed.

Synthesized control algorithms of car parking are realized in Python using popular mathematical libraries Matplotlib and NumPy. The computer verification of the synthesized algorithms based on the model of the SDC dynamics, which confirmed the effectiveness of the proposed algorithmic solutions, was carried out.

References

1. Kobylinsky, A.: Dangers and prospects of development of unmanned motor transport. Sci. Technol. Educ. **83**(3), 40–44 (2022)
2. Zhidkov, Ye.S., Shipovalov, D.A.: Development of the direction of unmanned transport: automatic car parking system. Internauka **47**(1), 6–8 (2019)
3. Nourinejad, M., Bahrami, S., Roorda, M.J.: Designing parking facilities for autonomous vehicles. Transport. Res. Part B Methodol. **109**(39), 110–127 (2018)
4. Singh, S., Saini, B.S.: Autonomous cars: recent developments, challenges, and possible solutions. IOP Conf. Series: Mater. Sci. Eng. **1022**(1), 012028 (2021)
5. Agrachev, A.A., Sachkov, Y.: Geometric Theory of Control [Geometricheskaya teoriya upravleniya]. Fizmatlit, Moscow (2005)
6. Vagizov, M.R., Khabarov, S.P.: Construction of software trajectories of motion on the basis of the solution of the problem "Dubins Machine." Inf. Space **3**, 116–125 (2021)
7. Mashtakov, A.P.: The problem of speed on a group of plane movements with control in a semicircle. Math. Coll. **213**(4), 100–122 (2022)

8. Zhdanov, A.A., Klimov, D.M., Korolev, V.V., Utemov, A.E.: Modeling of the process of parallel parking of a car. J. Comput. Syst. Sci. Int. **6**, 74–86 (2008)
9. Li, B., Shao, Z.: A unified motion planning method for parking an autonomous vehicle in the presence of irregularly placed obstacles. Knowl.-Based Syst. **86**, 11–20 (2015)
10. Gymez-Bravo, F., Cuesta, F., Ollero, A., Viguria, A.: Continuous curvature path generation based on b-spline curves for parking maneuvers. Rob. Auton. Syst. **56**(4), 360–372 (2008)
11. Vorobieva, H., Minoiu-Enache, N., Glaser, S., Mammar, S.: Geometric continuous curvature path planning for automatic parallel parking. In: Proceedings of 10th IEEE International Conference on Networking, Sensing and Control (ICNSC 2013), pp. 418–423 (2013)
12. Muller, B., Deutscher, J., Grodde, S.: Continuous curvature trajectory design and feedforward control for parking a car. IEEE Trans. Control Syst. Technol. **15**(3), 541–553 (2007)
13. Siedentop, C., Heinze, R., Kasper, D., Breuel, G., Stachniss, C.: Path-planning for autonomous parking with dubins curves. In: Proceedings of the Workshop Fahrerassistenzsysteme, pp. 1–8 (2015)
14. Ardentov, A.A., Gubanov, I.S.: Modeling of parking a car with a trailer along the Markov-Dubins and Reeds-Shepp paths. Program Systems: Theory and Applications **10**(4), 97–110 (2019)
15. Kong, S.-G., Kosko, B.: Comparison of Fuzzy, Neural Truck Backer Upper Control Systems. In: Proc. 1990-IJCNN International Joint Conference on Neural Networks, San Diego, CA, USA, vol. 3, pp. 349–358 (1990)
16. Li, T., Chang, S.: Autonomous Fuzzy Parking Control of a Car-Like Mobile Robot. IEEE Trans. Syst. Man Cybern. Part A: Syst. Hum. **33**(4), 451–465 (2003)
17. Zhao, Y., Collins, E.G., Jr.: Robust Automatic Parallel Parking in Tight Spaces via Fuzzy Logic. Robot. Auton. Syst. **51**(2), 111–127 (2005)
18. Mukeshimana, C.: Fuzzy Model of Parking Control for an Unmanned Vehicle. In: Proc. of International Scientific Conference on Control Problems in Technical Systems, vol. 1, pp. 432–436 (2017). (in Russian)
19. Ballinas, E., Montiel, O., Castillo, O., Rubio, Y., Aguilar, L.T.: Automatic parallel parking algorithm for a car-like robot using fuzzy PD+I control. Eng. Lett. **26**(4), 447–454 (2018)
20. Wang, Z., Shao, Q., Wang, C., Zhang, Q.: Automatic parking trajectory planning based on recurrent neural network. In: Proceedings of 2018 IEEE 9th International Conference on Software Engineering and Service Science (ICSESS), Beijing, China, pp. 1–4 (2018)
21. Parashar, S., Kumar, G.: Smart parking system using genetic optimization: a review. In: Proceedings of 2019 International Conference on Intelligent Sustainable Systems (ICISS), Palladam, India, pp. 599–603 (2019)
22. Evdokimova, T.S., Sinodkin, A.A., Fedosova, L.O., Tyrikov, M.I.: Algorithm for constructing a global trajectory of traffic and planning of the automatic parking route of the self-driving car. Vestnik MSTU «Stankin» **55**(4), 61–67 (2020)
23. Komarov, I., Lobach, D., Muthanna, A.S.A.: Intelligent parking control system for unmanned vehicles based on internet of things technologiesю. In: Proceedings of XI International Conference on Actual Problems of Infotelecommunications (APINO 2022), pp. 592–596 (2022)
24. Tyulenev, I.D., Filimonov, N.B.: Automatic control for parking self-driving car based on Dubins and Reeds-Shepp models. J. Adv. Res. Tech. Sci. **35**, 52–59 (2023)
25. Tyulenev, I.D., Filimonov, N.B.: Automatic parking control of an unmanned car based on reinforcement machine learning. High-Perf. Comput. Syst. Technol. **7**(1), 159–165 (2023)
26. Dubins, L.E.: On curves of minimal length with a constraint on average curvature, and with prescribed initial and terminal positions and tangents. Am. J. Math. **79**(3), 497–516 (1957)
27. Reeds, J.A., Shepp, L.A.: Optimal paths for a car that goes both forwards and backwards. Pac. J. Math. **145**(2), 367–393 (1990)

28. Kazakov, K.A., Semenov, V.A.: An overview of modern methods for motion planning. Proc. Inst. Syst. Program. RAS **28**(4), 241–294 (2016)
29. LaValle, S.M., Kuffner, J.J.: Rapidly-exploring random trees: progress and prospects. In: Proceedings of 2000 Workshop on the Algorithmic Foundations of Robotics, pp. 293–308 (2000)
30. Ma, Ts., Malinina, T.A., Borisik, M.M., Osipovich, V.S.: Machine learning algorithms car operation. In: Proceedings of 4th International Conference and Expo on Big Data Advanced Analytics, p. 416–418 (2018)
31. Dudakov, A.S., Tursunov, T.R., Filimonov, N.B.: The method of deep reinforcement learning in motion planning problem of mobile robots in an environment with obstacles. Mechatron. Autom. Rob. **11**, 7–13 (2023)
32. Moon, J., Bae, I., Kim, S.: Automatic parking controller with a twin artificial neural network architecture. Math. Prob. Eng. **2019**(6), 1–18 (2019)
33. Wang, Z., Shao, Q., Wang, C., Zhang, Q.: Automatic parking trajectory planning based on recurrent neural network. In: Proceedings of IEEE 9th International Conference on Software Engineering and Service Science (ICSESS), pp. 1–4 (2018)
34. Zhang, P., et al.: Reinforcement learning-based end-to-end parking for automatic parking system. Sensors **19**(18), 3996 (2019)
35. Kiran, B.R.: Deep reinforcement learning for autonomous driving: a survey. IEEE Trans. Intell. Transp. Syst. **23**(6), 4909–4926 (2021)
36. Jang, B., Kim, M., Harerimana, G., Kim, J.W.: Q-learning algorithms: a comprehensive classification and applications. IEEE Access **7**, 133653–133667 (2019)

Application of Smoothing Technique to Model Predictive Traffic Signal Control

Sergey V. Matrosov[1] and Nikolay B. Filimonov[1,2]([✉])

[1] Lomonosov Moscow State University, Leninskie Gory, 1/2, 119991 Moscow, Russia
matrosik14@gmail.com, nbfilimonov@mail.ru
[2] Bauman University, 2nd Baumanskaya str., 5/1, 105005 Moscow, Russia

Abstract. This paper introduces a new adaptive traffic signal control algorithm that operates within the model predictive control framework. Its primary objective is to enhance the traffic network's throughput under heavy load. To achieve this goal, a comprehensive traffic flow model and a specially tailored target function were used. The predictive model utilized in this algorithm is second-order macroscopic traffic model, enabling accurate prediction of traffic phenomena such as wave formations and nonlinear effects. Furthermore, the model can be refined using historical data, which enhances the precision of predictions. The paper outlines both the model itself and the numerical scheme utilized for its computations. The proposed target function considers the characteristics of a traffic dynamics and aims to provide a uniform distribution of vehicles in the transport network. Optimal control can be found as a solution to a continuous optimization problem related to a noisy zero-order oracle. The smoothing technique is used to solve the optimization problem. It allows using first-order stochastic optimization methods in the situation when the gradient of the target function is unknown. The developed traffic light control algorithm has been tested in the traffic simulation environment SUMO in the set of RESCO benchmarks.

Keywords: Transportation Network · Transport Modeling · Macroscopic Traffic Model · Control System · Traffic Signal Control · Model Predictive Control · Zero-order Optimization · Smoothing Technique

1 Introduction

The modern urban transportation infrastructure provides a wide variety of traffic monitoring and control tools. It allows remote control of traffic lights, collecting information obtained from detectors and cameras, monitoring the GPS tracks of service equipment, etc. At the same time, the problem of creating a traffic signal control system [1–3] providing the maximum throughput of the road network becomes increasingly important. The concept of adaptive traffic signal control is that the control of traffic lights is adjusted according to the current state of the transport network.

Many popular traffic control algorithms use simple traffic models [4–10]. One of the promising areas of development is to take into account the nonlinear dynamics of traffic flows in the control system. When the traffic is dense, vehicles begin to interact with each other, which leads to traffic flow instability [11], formation of traffic waves [12], and a hysteresis effect [13]. All this leads to a decrease in the transport network throughput.

The authors of this work develop further the traffic light control algorithm proposed in [14–16]. In contrast with the previous work, an optimal control problem was formulated as a continuous optimization problem concerning a noisy zero-order oracle. The smoothing technique [17] was used to solve this problem. It allows using stochastic first-order optimization algorithms in the proposed conditions.

2 Transport Network Model

2.1 Traffic Controllers

The traffic controller manages the traffic lights. It switches between several phases (see Fig. 1); each corresponds to a set of permitted directions of movement at intersection.

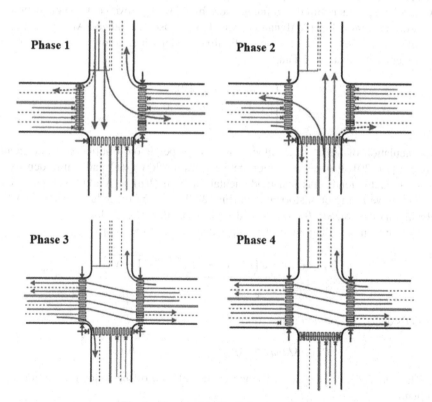

Fig. 1. Phases of traffic light controller.

The traffic controller will operate according to the installed program. It sets the sequence and the duration of phases within the traffic light cycle. When the program switches, the controller completes a current cycle and then switches to the new program.

We will assume that the phase sequence is fixed for each intersection. Let a vector of programs be $u = (u^1, \ldots, u^v)$ and the program for the controller v be $u^v = (\tau_k^v)_{1,\ldots,K^v}$, where τ_k^v is the duration of the phase k, and K^v is the number of phases of the controller v. The duration of each phase τ_k^v is within the interval of $\left[\tau_{min,k}^v, \tau_{max,k}^v\right]$. All the permitted vectors of the programs u form the convex set U.

2.2 Macroscopic Traffic Flow Model

The macroscopic approach is used to model the traffic flows behavior. The state of the road is described by the density $\rho(x, t)$, the average velocity $v(x, t)$, and the intensity $Q(x, t)$, where $x \in [0, L]$ is a position on the road and t is the current time. The traffic dynamics is defined by the system of partial differential equations.

The traffic network is represented as a directed weighted graph (L, C, A). Where $l \in L$ is a road in the network. Connection $c_{ij} \in C$ represents direction of movement at intersection from road i to road j. Weight $\alpha_{ij} \in A$ defines the turn-ratio from road i to road j. Controllers interact with the traffic network by allowing movement on connections.

In order to model traffic dynamics, we chose to use generalized ARZ model [18]. It is the second-order macroscopic traffic flow model which allows fine-tuning of its parameters using historical data.

$$\begin{cases} \rho_t + (\rho v)_x = 0; \\ (\rho\omega)_t + (\rho\omega v)_x = 0; \\ v = V(\rho, \omega). \end{cases}$$

Calculation of traffic flow velocity $V(\rho, \omega)$ depends on a notion of fundamental diagram [19, 20]. It gives a relation between traffic flow $Q(\rho)$ and traffic density ρ. GARZ model extends classical fundamental diagram $Q(\rho)$ with additional parameter ω, to improve fitting on historical data (Fig. 2). The fitting procedure for $Q(\rho, \omega)$ and calculation of average traffic flow speed $V(\rho, \omega)$ are described below.

The fundamental diagram curve is provided by the expression:

$$Q_{\alpha,\lambda,p}(\rho) := \alpha\left(a + (b - a)\frac{\rho}{\rho_{max}} - \sqrt{1 - y^2}\right);$$
$$a = \sqrt{1 + \lambda^2 p^2}; \; b = \sqrt{1 + \lambda^2(1 - p)^2}; \; y = \lambda(\rho/\rho_{max} - p).$$

Let's denote historical data as follows:

$$(\rho_{data}, Q_{data}) = \{(\rho_i, Q_i) : i = 1, \ldots, N_{data}\}.$$

The parameters α, λ, p are obtained as the solution of the following optimization problem:

$$argmin_{\alpha,\lambda,p}\left(\beta\|Q_{\alpha,\lambda,p}(\rho_{data}) - Q_{data}\|_+^2 + (1 - \beta)\|Q_{\alpha,\lambda,p}(\rho_{data}) - Q_{data}\|_-^2\right)$$
$$\|Q_{\alpha,\lambda,p}(\rho_{data}) - Q_{data}\|_+^2 = \sum_{i=1}^{N_{data}} max\{Q_{\alpha,\lambda,p}(\rho_i) - Q_i, 0\}$$
$$\|Q_{\alpha,\lambda,p}(\rho_{data}) - Q_{data}\|_-^2 = \sum_{i=1}^{N_{data}} max\{Q_i - Q_{\alpha,\lambda,p}(\rho_i), 0\}$$

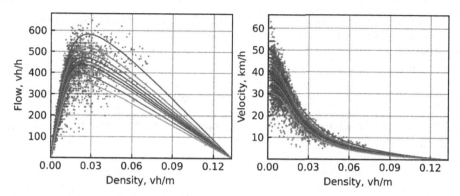

Fig. 2. Data-fitted fundamental diagram plotted for the GARZ model.

The parameter β defines the proportion of historical points located under the curve $Q_{\alpha,\lambda,p}(\rho)$. Solving the optimization problem for different $\beta_i \in (0, 1)$, we can obtain a collection of points $\alpha(\beta_i)$, $\lambda(\beta_i)$, $p(\beta_i)$ and a family of functions $Q(\rho, \beta_i)$.

Let us consider that ω is the maximum speed of the driver, i.e. $\omega = V(0, \omega)$. Then we can use relation $Q = \rho V$ and construct a mapping from β to ω:

$$V(\rho, \beta) = \begin{cases} Q_\beta(\rho)/\rho, \rho > 0 \\ Q'_\beta(\rho), \rho = 0 \end{cases}; \quad \omega = V(0, \beta) = \frac{\alpha_\beta}{\rho_{max}}\left(b_\beta - a_\beta + \frac{\lambda_\beta^2 p_\beta}{a_\beta}\right).$$

Now we can calculate $\omega_i = \omega(\beta_i)$ and fit polynomial regression on $\alpha(\omega_i)$, $\lambda(\omega_i)$ and $p(\omega_i)$ for $\omega \in [\omega_{min}, \omega_{max}]$. Hence, we obtain the functions $\alpha(\omega)$, $\lambda(\omega)$, $p(\omega)$ and can calculate $Q(\rho, \omega)$ and $V(\rho, \omega)$ in the domain $[0, \rho_{max}] \times [\omega_{min}, \omega_{max}]$. We can extend this domain to $[0, \rho_{max}] \times R_+$ projecting ω on the segment $[\omega_{min}, \omega_{max}]$.

2.3 Numerical Scheme

Godunov's method is used to find an approximate solution. The computational domain is divided into volumes, and averaged traffic parameters are assigned to each cell. The finite-difference scheme for the above system is written as follows:

$$\rho_i^{n+1} = \rho_i^n - \frac{\Delta t}{\Delta x}\left(Q_{i+1/2}^n - Q_{i-1/2}^n\right);$$
$$y_i^{n+1} = y_i^n - \frac{\Delta t}{\Delta x}\left(\omega_{i-1}^n Q_{i+1/2}^n - \omega_i^n Q_{i-1/2}^n\right);$$
$$\omega_i^n = y_i^n / \rho_i^n.$$

To calculate the flow $Q_{I\pm1/2}^n$ between two cells, we need to solve Riemann problem. For numerical computation of this flow, we use formalism of sending and receiving functions proposed in [21].

Let's denote cell on the left side of the boundary as $U_L = (\rho_L, \omega_L)$, and cell on the right side as $U_R = (\rho_R, \omega_R)$. Sending function $S(U_L, U_R)$ and receiving function $R(U_L, U_R)$ define maximum flow which U_L could sent and U_R could receive. Therefore, resulting flow between cells is given by expression $Q = min\{S, R\}$.

A correct solution of the Riemann problem in the case of the GARZ model requires introducing an intermediate state $U_M = (\rho_M, \omega_M)$ between the cells. In the case of $\omega_L < v_R = V(\rho_R, \omega_R)$, the intermediate state is $U_M = (0, \omega_L)$. Otherwise, it is calculated from the equations:

$$\rho_M : u_R = V(\rho_M, \omega_L); \ \omega_M = \omega_L.$$

To calculate the sending and receiving functions, we need to know critical density and maximum flow:

$$\rho_c(\omega) = argmax_\rho Q(\rho, \omega); \ Q^{max}(\omega) = max_\rho Q(\rho, \omega).$$

Then the sending and receiving functions applied for the GARZ model are specified as:

$$S(\rho_L, \omega_L) = \begin{cases} \rho_L \omega_L, & \rho_L \leq \rho_c(\omega_L) \\ Q^{max}(\omega_L) & \rho_L > \rho_c(\omega_L) \end{cases}; \ R(\rho_M, \omega_L) = \begin{cases} Q^{max}(\omega_L) & \rho_M \leq \rho_c(\omega_L) \\ \rho_M \omega_L & \rho_M > \rho_c(\omega_L) \end{cases}.$$

2.4 Intersection Model

Modelling the traffic dynamics at an intersection necessitates describing the way the traffic flows splits and merge. In order to do so, we extend the usage of sending and receiving functions for intersection. We will divide total flow S_i of the last cell of the road i to sub-flows S_{ij} for each direction c_{ij}. S_{ij} describes maximum possible flow from road i to road j and calculates as follows $S_{ij} = \alpha_{ij} \delta_{ij} S_i.$, where δ_{ij} indicates that movement in this direction is allowed, and α_{ij} is a turn-ratio for this road in intersection.

For traffic splitting calculations are straightforward. For each direction c_{ij} we will apply standard procedure described above, where U_L is the last cell of road i and U_R is first cell of road j, but value of sending function is equal to S_{ij}.

We assume that when M traffic streams merge drivers strive to maximize total flow $Q_j = \sum_{i=1}^{M} q_{ij}$ to road j. Considering constraints imposed by S_{ij} and R_j we could formulate traffic flows merge as following optimization problem:

$$max_{q_{ij}} \left(\sum_{i=1}^{M} q_{ij} \right);$$
$$0 \leq q_{ij} \leq S_{ij}, \forall i;$$
$$\sum_{i=1}^{M} q_{ij} \leq R(\rho_{j-}, \omega_{j-}).$$

To calculate the receiving function R for the road j, we have to find the intermediate state of the flow at the entrance $U_{j-} = (\rho_{j-}, \omega_{j-})$. We will use following equations:

$$\omega_{j-} = \left(\sum_{i=1}^{M} \omega_i q_{ij} \right) / \left(\sum_{i=1}^{M} q_{ij} \right);$$
$$v_{j-} = min\{\omega_{j-}, v_j\};$$
$$\rho_{j-} : v_{j-} = V(\rho_{j-}, \omega_{j-}).$$

We assume that all directions have equal priority when traffic streams merge to road j and drivers do not cooperate to pass intersection. Then resulting flows q_{ij} should be proportional to possible maximum flows S_{ij}:

$$q_{ij} = \beta_{ij} Q_j; \; \beta_{ij} = S_{ij} / \left(\sum_{i=1}^{M} S_{ij} \right).$$

After substituting the expression for q_{ij} into the original optimization problem, we can calculate the exact solution of a merge problem. Total flow could be calculated as follows:

$$Q_j = \min \left\{ \sum_{i=1}^{M} S_{ij}, R_j \right\}.$$

3 Model Predictive Control

We will update programs on controllers at discrete time moments $t_i = i \Delta T$. It is assumed that we know the current state of the transport network s_i and the input flows d_i at a time t_i. Then we can make predictions of system behavior for the chosen control u_i within some time horizon T using model of transport network F. These predictions can be used to compare performance of different controls u via evaluation of target function $J_F(s_i, d_i, u)$. The optimal control is then determined by solving the following optimization problem:

$$argmin_{u_i \in U} J_F(s_i, d_i, u_i).$$

In this paper, we use a specially designed target function that estimates the risk of an excessive traffic concentration on one part of the transport network.

3.1 Target Function

A more uniform load distribution is assumed to increase the average road capacity and to reduce the probability of traffic jam buildup. To implement this strategy, the following target function is used:

$$CDS(s_i, d_i, u_i) = \frac{1}{T} \sum_{t=0}^{T} \sum_{l \in L} w(l) \cdot \left(\frac{\overline{\rho}_l^t}{\rho_c(\omega_{max})} \right)^2; \; \overline{\rho}_l^t = \frac{1}{M_l} \sum_{m=1}^{M_l} \rho_m^t;$$

$$WPS(s_i, d_i, u_i) = \frac{2}{T^2(T-1)} \sum_{t=0}^{T} \sum_{c \in C} t_{closed}^c \left(\overline{\rho}_{c_{in}}^t - \overline{\rho}_{c_{out}}^t \right);$$

$$J_F(s_i, d_i, u_i) = CDS(s_i, d_i, u_i) + WPS(s_i, d_i, u_i).$$

$CDS(s_i, d_i, u_i)$ (critical density score) reflects the accumulation of the penalty for excessive density on the road in the transport network. We can estimate the level of congestion of the road $l \in L$ comparing average density $\overline{\rho}_l^t$ with critical density ρ_c. Then we square resulting ratio to make penalty more severe when traffic density on the road exceeds critical density. Weight of the road $w(l)$ can be used to increase the priority of certain sections of the transport network. We assume than long sections are more likely to generate traffic waves so by default, the weight is set proportional to the road length.

$WPS(s_i, d_i, u_i)$ (weighted pressure score) is designed to penalize the inefficient phase utilization. The set C contains all controllable directions in the traffic network. Let us estimate t^c_{closed} for each direction $c \in C$. When the direction c is open (traffic movement is allowed in this direction), $t^c_{closed} = 0$; otherwise, it shows how long this direction was closed. In the case of the closed direction c, let us estimate the traffic pressure, which equals $\overline{\rho}^t_{C_{in}} - \overline{\rho}^t_{C_{out}}$ where $\overline{\rho}^t_{C_{in}}$ and $\overline{\rho}^t_{C_{out}}$ are average densities on the incoming and outgoing road at the moment t. The high traffic pressure indicates that there is a high demand for movement in this direction and it could be fulfilled. Hence, this metrics should penalize unnecessary long traffic light cycles and give priority to the busiest directions.

3.2 Zero-Order Smoothing Optimization Technique

The target function is calculated based on the simulation results, so it is impossible to find its gradient. Let us fix the current state of the transport network s_i and the predicted input flows d_i and denote by $f(u) = J_F(s_i, d_i, u)$. The numerical study of the target function shows that it is a non-smooth function, which can be represented as a smooth function with noise added to it. Then the search for optimal control is formulated as an optimization problem having a noisy zero-order oracle:

$$\text{argmin}_{u_i \in U} f(u).$$

In [17] the method that allows using first-order stochastic optimization algorithms for this class of problems was proposed and its convergence rates were estimated.

We propose to replace the target function $f(u)$ with its smoothed counterpart $f_\gamma(u)$. This is done by averaging over an euclidean sphere of the radius γ:

$$f_\gamma(u) = E_r f(u + \gamma r),$$

where the random vector r is uniformly distributed on the unit euclidean sphere. If the function $f(u)$ is Lipschitz, then the function $f_\gamma(u)$ and its gradient are shown to be also Lipschitz.

The following unbiased estimate with bounded variance is used to approximate the gradient of the function $f_\gamma(u)$:

$$\nabla f_\gamma(u, e) = d\frac{f(u + \gamma e) - f(u - \gamma e)}{2\gamma} e,$$

where d is the dimensionality of the space U, and e is a random vector uniformly distributed on a unit sphere.

The original problem can be replaced by the following one:

$$\text{argmin}_{u_i \in U} f_\gamma(u).$$

In this case, the function $f_\gamma(u)$ has a Lipschitz gradient and the approximation $\nabla f_\gamma(u, e)$ is an unbiased stochastic gradient having a small variance, which allows applying stochastic optimization methods of the first order. The work [17] shows that the $\epsilon/2$-solution of the smoothed problem will also be the ϵ-solution of the original problem.

4 Numerical Experiments

To test the proposed control algorithm, we used the "Simulation of Urban MObility" (SUMO) [22] software. It is an open source, highly portable, microscopic and continuous traffic simulation package designed to handle large networks. In the framework of the experiment, the real transport network data was substituted by the synthetic data generated by SUMO.

Benchmark control problems were taken from the RESCO benchmark set [23]. It includes two synthetic (Arterial4x4 and Grid4x4) and several realistic traffic scenarios (Fig. 3). The traci utility was used to interact with the SUMO environment (data collection and traffic lights control). Several minor modifications were made for the traffic scenarios, and they are listed below.

We added road detectors (inductive loops) to the traffic networks to collect data for the traffic model validation. Three traffic detectors (at the beginning, in the middle, and at the end) were installed on each lane.

In order to convert the traffic network graphs from the SUMO format to the format that is appropriate for the proposed macroscopic model, we marked junctions on the boundary of the simulated traffic network. An additional markup allowed conducting a conversion between the formats programmatically.

Weights α_{ij} were calculated based on the vehicles routes statistics. We collected this data with standard SUMO tools during preparatory runs of the test traffic scenarios.

There are several different types of roads having different speed limits in the test graphs. Fundamental diagrams were made for each type, but these fundamental diagrams may be shared between different traffic scenarios. We used a common scenario to generate the fundamental diagram; we gradually increased the traffic density in a circular route involving the road of a particular type and the data collected from the loop detectors.

The scripts developed for the graph conversion and data preparation can be found in this Github repository [24].

To validate the accuracy of the traffic model, we collected detectors data during SUMO simulation run. Then, we replicated this simulation with macroscopic model and recorded traffic parameters. For each detector we estimated RMSE error of the model prediction at the detector location. Then we calculated average error by averaging error over all non-empty lanes. Estimated errors are provided in Table 1. The relative error is calculated using the maximum values specified by the fundamental diagram for the corresponding quantities and is shown in parentheses.

Table 2 shows the results of the proposed control algorithm and their comparison with fixed plans and the genetic algorithm described in [16]. We used a control cycle duration of 300 s having a prediction horizon of 450 s, and a stochastic gradient descent was used as the basis for the optimization algorithm. The results of the smoothing method

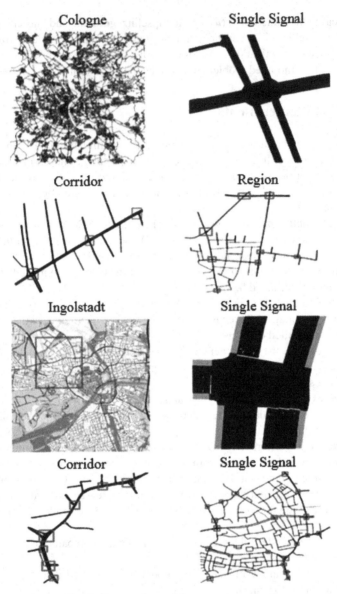

Fig. 3. Test sections of the transport network.

were averaged after 5 runs. In the case of the Cologne1 and Ingolstadt 21 scenarios, the algorithm showed unsatisfactory performance: there was a significant variation between the target function values across runs, and in some cases, there was a minor drop in throughput. For other scenarios, the results are shown in Table 2.

Table 1. Model error (RMSE) compared to the SUMO simulation.

	Intensity	Velocity	Density
Arterial4x4	169.8 vh/h (7.9%)	13.8 km/h (25.1%)	0.043 vh/m (21.4%)
Grid4x4	59.9 vh/h (2.8%)	5.0 km/h (9.1%)	0.001 vh/m (0.4%)
Cologne1	194.0 vh/h (9.0%)	16.7 km/h (30.4%)	0.031 vh/m (15.6%)
Cologne3	126.1 vh/h (5.8%)	11.4 km/h (20.7%)	0.015 vh/m (7.3%)
Cologne8	122.7 vh/h (5.7%)	8.5 km/h (15.4%)	0.012 vh/m (5.7%)
Ingolstadt1	210.7 vh/h (9.8%)	12.2 km/h (22.3%)	0.029 vh/m (14.5%)
Ingolstadt7	154.7 vh/h (7.2%)	14.3 km/h (26.1%)	0.030 vh/m (14.8%)
Ingolstadt21	166.8 vh/h (7.7%)	12.8 km/h (23.3%)	0.013 vh/m (6.3%)

Table 2. Performance of the control methods.

	Fixed program	Genetic Algorithm	Smoothing Technique
Arterial 4 × 4			
Delay	647.5s	–	515.6 s (20.4%)
Queue	16.97 m	–	16.17 m (4.7%)
Grid 4 × 4			
Delay	90.2 s	–	89.5 s (0.7%)
Queue	0.71 m	–	0.70 m (0.49%)
Cologne 3			
Delay	35.7 s	31.23 s (5.28%)	31.9 s (10.6%)
Queue	1.43 m	1.15 m (8.51%)	1.17 m (17.7%)
Cologne 8			
Delay	43.0 s	38.94 s (8.83%)	39.0 s (9.4%)
Queue	0.56 m	0.46 m (16.51%)	0.48 m (15.3%)
Ingolstadt 1			
Delay	27.9 s	–	20.1 s (28.0%)
Queue	2.85 m	–	1.53 m (46.4%)
Ingolstadt 7			
Delay	61.8 s	56.36 s (8.83%)	52.5 s (15.1%)
Queue	1.41 m	1.25 m (11.61%)	1.07 m (24.3%)

5 Conclusion

The proposed traffic signal control algorithm showed its consistency in synthetic tests. The transport network model predicts traffic behavior with reasonable accuracy. The smoothing method was used to optimize the target function, which allows applying stochastic first-order optimization algorithms in the problems having a noisy zero-order oracle. The algorithm has proved its consistency being used in a computer experiment. Both synthetic and real scenarios were used for testing.

It is important to note the flexibility of the proposed approach. The control algorithm makes it possible to quickly replace the transportation network model, the target function, or the basic stochastic optimization algorithm. Each of these options sets a potential direction for the development.

References

1. Bretherton, R.D.: Scoot urban traffic control system-Philosophy and evaluation. IFAC Proc. Vol. **23**(2), 237–239 (1990)
2. Samadi, S., Rad, A.P., Kazemi, F.M., Jafarian, H.: Performance evaluation of intelligent adaptive traffic control systems: a case study. J. Transport. Technol. **2**(3), 248 (2012)
3. Khattak, Z.H., Magalotti, M.J., Fontaine, M.D.: Operational performance evaluation of adaptive traffic control systems: a bayesian modeling approach using real-world GPS and private sector PROBE data. J. Intell. Transport. Syst. **24**(2), 156–170 (2020)
4. Varaiya, P.: Max pressure control of a network of signalized intersections. Transport. Res. Part C: Emerg. Technol. **36**, 177–195 (2013)
5. Lioris, J., Kurzhanskiy, A., Varaiya, P.: Adaptive max pressure control of network of signalized intersections. IFAC-PapersOnLine **49**(22), 19–24 (2016)
6. Diakaki, C., Papageorgiou, M., Aboudolas, K.: A multivariable regulator approach to traffic-responsive network-wide signal control. Control. Eng. Pract. **10**(2), 183–195 (2002)
7. de Oliveira, L.B., Camponogara, E.: Predictive control for urban traffic networks: initial evaluation. IFAC Proc. Vol. **40**(2), 424–429 (2007)
8. Aboudolas, K., Papageorgiou, M., Kouvelas, A., Kosmatopoulos, E.: A rolling-horizon quadratic-programming approach to the signal control problem in large-scale congested urban road networks. Transport. Res. Part C: Emerg. Technol. **18**(5), 680–694 (2010)
9. Lin, S., Schutter, B.D., Xi, Y., Hellendoom, H.: Fast model predictive control for urban road networks via MILP. IEEE Trans. Intell. Transp. Syst. **12**(3), 846–856 (2011)
10. Lin, S., Schutter, B.D., Hellendoom, H.: Efficient network-wide model-based predictive control for urban traffic networks. Transport. Res. Part C: Emerg. Technol. **24**, 122–140 (2012)
11. Edie, L.C.: Car-following and steady-state theory for non-congested traffic. Oper. Res. **9**(1), 66–76 (1961)
12. Newell, G.F.: Instability in dense highway traffic: a review. In: Proceedings the 2nd International Symposium of on the Theory of Traffic Flow (1963)
13. Treiterer, J., Myers, J.: The hysteresis phenomenon in traffic flow. Transport. Traffic Theory **6**, 13–38 (1974)
14. Matrosov, S.V.: The concept of predictive traffic signal control with trainable model of transport network. J. Adv. Res. Tech. Sci. **25**, 56–62 (2021)
15. Matrosov, S.V.: Algorithm prognostic control of system of crossroads based on macroscopic models of traffic flows. J. Adv. Res. Techn. Sci. **27**, 80–87 (2021)

16. Matrosov, S.V., Filimonov, N.B.: Adaptive traffic signal control based on a macroscopic model of the transport network. Commun. Comput. Inf. Sci. **1773**, 219–229 (2023)
17. Gasnikov, A., et al.: The power of first-order smooth optimization for black-box non-smooth problems. Proc. Mach. Learn. Res. **162**, 7241–7265 (2022)
18. Fan, S., Herty, M., Seibold, B.: Comparative model accuracy of a data-fitted generalized Aw-Rascle-Zhang model. Netw. Heterogeneous Media **9**(2), 239–268 (2014)
19. Greenshields, B.D., Thompson, J.T., Dickinson, H.S., Swinton, R.S.: The photographic method of studying traffic behavior. In: Highway Research Board Proceedings, vol. 13 (1934)
20. Greenshields, B.D., Bibbins, J.R., Channing, W.S., Miller, H.H.: A study of traffic capacity. In: Highway Research Board Proceedings, vol. 1935. National Research Council (USA), Highway Research Board (1935)
21. Lebacque, J.-P., Haj-Salem, H., Mammar, S.: Second order traffic flow modeling: supply-demand analysis of the inhomogeneous Riemann problem and of boundary conditions. In: Proceedings of the 10th Euro Working Group on Transportation (EWGT), vol. 3 (2005)
22. Lopez, P.A. et al.: Microscopic Traffic Simulation using SUMO. In: Proceedings of 21st International Conference on Intelligent Transportation Systems (ITSC), pp. 2575–2582 (2018)
23. Ault, J., Sharon, G.: Reinforcement learning benchmarks for traffic signal control. In: Proc. of Thirty-Fifth Conference on Neural Information Processing Systems Datasets and Benchmarks Track (Round 1) (2021)
24. Matrosov, S.V.: Gradient-free optimization in the traffic light control problem. https://github.com/matrosik17/sirius_dfo_traffic_control. Accessed 14 Aug 2023

Design Features of the Frequency-Controlled Electric Drive for Positioning Mechanisms

Ishembek Kadyrov$^{(\boxtimes)}$ ⓘ, Baktybek Turusbekov, Bermet Zhanybekova, and Baktybek uulu Azamat

Skryabin Kyrgyz National Agrarian University, Mederov Str. 68, 720005 Bishkek, Kyrgyz Republic

bgtu_kg@mail.ru

Abstract. The paper is devoted to the design peculiarities of the frequency-controlled electric drive of the hydraulic distributor of the ultra-high pressure hydraulic press. It justifies the need to power the windings of an induction motor (IM) with a short-circuit rotor from the output of a direct frequency converter (DFC) assembled according to a symmetrical scheme using six-pulse or three-pulse thyristor transducers. The paper also shows the method of the electric drive synthesis for positioning mechanisms, including the press hydraulic distributor, as well as the need to build a three-circuit system of subordinated coordinate control for the frequency-controlled electric drive of the hydraulic distributor. Unlike positioning DC electric drives the synthesis of the torque inner circuit in a frequency-controlled electric drive is significantly different, the features of which are described in detail in the paper. As a result of the measures taken, the required dynamic characteristics of the electric drive are provided in the torque circuit, the required processing time of the alignment is ensured in the speed loop, and the required accurate positioning of the working element is achieved in the position circuit. The study demonstrates the method when currents are generated in the AC stator windings with adjustable frequency, amplitude and phase by supplying sine signals from the output of the microprocessor generator to the inputs of the control system of the DFC power units. The paper shows the adjustment method, when the input of the torque regulator of the positive feedback, as well as its critical setting, is provided to the electric drive of the torque source property. The results of the synthesis of the electric drive control system are confirmed by a widely proposed experiment on the industrial model of the hydraulic distributor of the 30,000-ton press. Practical recommendations are formulated for the selection of rational power circuit layout diagrams of the direct frequency converter-induction motor system, as well as the principles for building an electric drive control system, which can be used in the design of the automation process with the participation of AC electric drives of positioning mechanisms.

Keywords: Hydraulic Distributor · Direct Frequency Converter · Induction Motor · Tachometer Generator · Thyristor Converter · Logarithmic Amplitude-Frequency Characteristic

V. Jordan et al. (Eds.): HPCST 2023, CCIS 1986, pp. 222–233, 2024.
https://doi.org/10.1007/978-3-031-51057-1_17

1   Introduction

According to their function, the positioning electric drives shall ensure processing of the specified movement of the actuator with the required accuracy and speed, with reliable limitation of the maximum torque.

The studied three-cylinder hydraulic press has a hydraulic circuit in which the working element moves with the help of a liquid in a hydraulic cylinder in the form of mineral oil pumped using high-speed pumps. The movement of the slide of the hydraulic press of super-high pressures can be attributed to positional ones, since in the process of performing a production operation, for example, cold pressing, the slide moves in cycles set by the machine-building engineers. At the same time, gear drives of hydraulic distributors of high-power presses must handle the maximum mismatch for a time not exceeding 2 s with an accuracy of not more than 3% [1, 2].

2   Problems of Positioning Mechanism Automation

Frequent transient modes of the actuator definitely determine the use of a fully controlled electric drive with high static and dynamic characteristics. Currently, AC and DC are almost equal in their adjustment characteristics, and the obvious advantages of AC machines compared to DC motors determine the feasibility of using an asynchronous electric drive [3].

When choosing a hydraulic distributor electric drive system, the factor of slow movement of the working element in the metal forming cycle is taken into account, therefore, the developers prefer the use of a frequency-controlled electric drive, in which the stator windings of the induction motor are powered from the output of the direct frequency converter (DFC).

Instead of a traditional frequency converter with a DC block, the choice of the DFC is driven by the noticeable advantage of the latter, firstly, due to a single conversion of electric energy, and secondly, due to the ease of generating an output voltage in which a smooth component of the load current is close to sinusoidal.

The DFC power circuit is assembled using reversible thyristor converters, the number of which, with a symmetrical circuit, should be equal to the number of phases of the induction motor. At the same time, the generation of the current in the DFC close to sinusoidal is achieved the more accurately, the lower the speed of rotation of the shaft of the induction motor. Besides, the use of a symmetrical DFC circuit consisting of three reversible thyristor generators (TG) is quite attractive since the electric drive retains its controllability even if one of the TG sets fails, which is proved by the studies of incomplete-phase, asymmetric DFC circuits [3].

2.1   Control Panel Functions

The electric drive according to the DFC-IM system in relation to the hydraulic distribution of the press should have a three-circuit system of subordinate coordinate control. At the same time, the inner circuit is the torque circuit, the optimal adjustment of which allows achieving the required dynamic characteristics of the electric drive. The next contour is a speed loop, which provides the required processing time for the mismatch, the speed drawdown may be significant. Finally, the outer loop is the position contour necessary for accurate positioning of the working element.

When choosing a structural scheme for controlling the electric drive according to the DFC-IM system, the task was to provide the same properties of the frequency-controlled electric drive as those of a DC electric drive. This task is quite difficult, since the main difference in a DC motor is the presence of two independent control channels: flow constancy is maintained through the excitation winding circuit in a simple way; electric energy is converted to mechanical energy via the anchor chain. In an induction motor with a short-circuited rotor, electric energy is supplied only to the stator windings. In this case, part of the electric energy in the stator winding is consumed to create a rotating magnetic field of the engine, and the remaining energy is transformed into rotor windings and converted into mechanical. Maintaining the flux linkage of the rotor at a constant level by simple means independent of the load is a task that is solved in this study.

2.2 Design Constraints

The experience in designing frequency-controlled electric drives for ditching machines showed that the simplest asynchronous electric drive systems can be obtained on the basis of frequency-current control using the compensation principle of torque control by absolute sliding. A prerequisite for this system is the constant flow of the rotor, which can be achieved if the known ratios between the amplitude, phase and frequency of the stator field are satisfied [3, 4].

In order to determine these ratios, let us write the equations of IM variables using the model of a two-phase generalized machine [5, 6]. In a two-phase synchronous coordinate system xy, for IM torque control at $\overline{\Psi}_2 = const$ it is necessary to ensure the formation of a stator current vector in the form of:

$$\bar{i}_1 = I_{1\,max} e^{j(\omega_{0el}t + \varphi)}, \tag{1}$$

where I_{1max} – current amplitude; φ – angle formed by the inductive load.

The amplitude value and angle of the generated current are determined by the following ratios:

$$I_{1\,max} = \Psi_{2\,max} / L_{12} \cdot \sqrt{1 + (L_2 \cdot S_a \cdot \omega_{0eln} / R_2')}, \tag{2}$$

$$\varphi = arctg\left((L_2 / R_2') S_a \omega_{0eln}\right), \tag{3}$$

where S_a – absolute slip.

Equations (1–3) can be solved using microprocessor means where by selecting the appropriate software it is possible to achieve conditions in which the flux linkage of the rotor will be maintained at a constant level over the entire range of speed control at any loads. It should be emphasized that this method of maintaining the flux linkage of the rotor at a constant level by generating the phase currents of the induction motor according to Eqs. (1–3) is indirect. Hence, it is necessary to conduct experimental studies for the reliability of this statement [7].

In order to clarify the validity of using an indirect method of maintaining the flux linkage, let us consider the vector current diagram shown in Fig. 1. The vector diagram is built on a complex plane without taking into account the active component of reactance in the magnetization circuit and the reactive component of reactance in the rotor circuit due to their insignificant values.

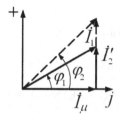

Fig. 1. Vector diagram of AM phase currents.

The diagram in Fig. 1 clearly explains the process of maintaining the magnetization current I_μ = const when the load on the motor shaft increases. At any induction motor speed, an increase in load causes an increase in rotor current $\dot{I}'_2$. In the absence of a frequency control system according to Eqs. (1–3), the demagnetizing effect of the rotor current fully appears and the magnetizing current of the machine with the growth of the rotor current can quickly decrease. The use of a microprocessor sinusoidal signal generator (MSSG) allows maintaining the flux linkage of the rotor Ψ_2 = const. Figure 1 shows that the increased rotor current gain $\dot{I}'_2$ is compensated by the phase control φ so that the increase in stator current I_1 results in load balancing only by the developed engine torque [8, 9].

Generation of currents in the DFC operating as a current source (1) requires the voltage applied to the current of an induction motor phase, for example, phase A using the following equation:

$$u_a = U_m \sin(\omega t + \varphi), \tag{4}$$

where φ – adjustable parameter.

In order to avoid the mathematical effect of summing the arguments (4), the expression (4) should be decomposed into the following components:

$$u_a = U'_m \sin \omega t + U_c \cos \omega t, \tag{5}$$

Then the amplitude module of the setting voltage (4) is determined by the following difference:

$$U_m = \sqrt{(U'_m)^2 + (U_c)^2}, \tag{6}$$

and the phase of the setting voltage is calculated using the following expression:

$$\varphi = arctg\left(U_c / U'_m\right). \tag{7}$$

According to Eq. (5), the constant flux linkage of the rotor $\overline{\Psi}_2 = const$ is ensured by maintaining the voltage amplitude U_c at a given level. In this case, both the amplitude (6) and phase setting the voltage (7) can be regulated due to change of one parameter U'_m, and the voltage on current amplitude dependence $u_{rc} \equiv U'_m$ on voltage torque setting u_{rt} will be linear.

Based on Eqs. (4–7), the MSSG shall generate current signals by solving equations of the form [9]:

$$\begin{cases} u_{rcA} = U'_m \sin \omega_0 t + U_c \cos \omega_0 t, \\ u_{rcB} = U'_m \sin(\omega_0 t + 120°) + U_c \cos(\omega_0 t + 120°), \\ u_{rcC} = U'_m \sin(\omega_0 t - 120°) + U_c \cos(\omega_0 t - 120°). \end{cases} \quad (8)$$

The block diagram of the MSSG model based on Eqs. (8) is in Fig. 2.

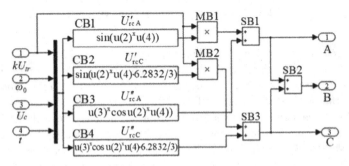

Fig. 2. Block diagram of the model of the sinusoidal signal generator.

The computing block SV1 in Fig. 1 generates a sinusoidal function of the first term of Eq. (8). The amplitude value for this component is adjusted at the input of the MV1 multiplication, where a voltage proportional to the regulated IM torque kU_{tr} is applied. At the same time, the product of the amplitude value with the generated SV1 sinusoidal function forms the first term of the current setting voltage. The voltage U_c maintains the amplitude value of the IM magnetizing current at a constant level in the SV3 block, which is multiplied with the cosine varying signal and forms the second term of the current setting. The calculation result of the SV3 block is added to the calculation result of the SV1 block at the input of the SV1 summing block. According to the similar algorithm, the MSSG generates a current setting voltage for phase C using the blocks SV2, SV4 and SV3.

The second MSSG Eq. (8) is calculated by simply summing the signals from the outputs of SV1 and SV3 using the adder SV2, as shown in Fig. 2.

The functional diagram of the integrated positioning motor of the press is shown in Fig. 3. The signal $U_{\Delta r}$ proportional to the mismatch of the set and actual position of the actuating element (AE), which is brought into double motion by the engine through the reducer P, generates a phase-sensitive detector, to the input of which signals are received from the selsyns BS and BR, which are proportional to the set and actual rotation angle of the AE. Further, this signal is supplied to the input of the proportional position regulator, the output signal of which is a task for the speed of the electric drive. The position controller is designed so that at high mismatch angles the drive operates in accordance with the algorithm included in the speed controller [7, 10].

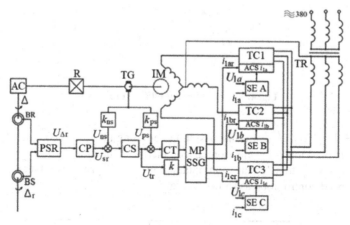

Fig. 3. Functional diagram of positioning frequency-controlled hydraulic distributor electric drive.

In the diagram in Fig. 2, the variable input signals for MSSG are a signal for setting the frequency ω_0 taken from the output of the torque regulator (TR) (Fig. 3); signal for setting the amplitude kU_{tr}, which is taken from the output of the speed controller (SC). The constant input signal is U_c, which is set once before the start-up and maintained at a constant level during the MSSG operation, since this parameter sets the magnetizing current in the IM. The parameter t is the current variable for the calculation blocks SV1, SV2, SV3, SV4 in which the harmonically varying terms of the 1st and 3rd equations of the system (8) are calculated.

2.3 Principle of Operation

The results of calculation of MSSG setting voltages in Fig. 4 are shown in the form of oscillograms of phase current setting signals at the reverse of the electric drive according to the DFC-IM system. The need to use a positive speed ratio shown in Fig. 3 is explained by the complexity of measuring the torque on the motor shaft by available means, and the indirect method of extracting the signal value proportional to the electromagnetic torque of the motor is associated with the complexity of calculating the equations of electromechanical energy conversion.

The fulfillment of the conditions described by Eqs. (1, 2, 3) and the introduction of the torque of the positive speed feedback with its critical setting at the input of the regulator allows presenting the transfer function of the optimized torque circuit in the form of an aperiodic block:

$$W_t(p) = \frac{k_t}{1 + T_{el}p},$$

(9)

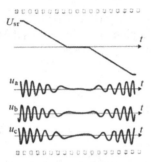

Fig. 4. Oscillograms of DFC-IM reverser.

where $k_t = \Psi^2_{2\max} \cdot P_p / R'_2$ – transmission factor along the torque contour, T_{el} – small time constant determined by electromagnetic inertia of the motor.

The torque loop transfer function obtained in Eq. (9) provides a block diagram of the motor speed control loop as shown in Fig. 5.

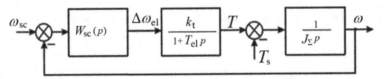

Fig. 5. Block diagram of electric drive speed control circuit.

According to the block diagram in Fig. 5, the transfer function of the speed control object will be as follows:

$$W_{sro} = \frac{k_t}{(1 + T_{el}p)J_\Sigma p}.$$

(10)

3 Method of Adjusting the Speed of a Positioning Mechanism

Stepwise load changes characteristic of hydraulic distributor mechanisms are a feature of the electric drive, which determine the need to constantly maintain the engine rotation speed at a given level with great accuracy. These features are taken into account in the synthesis of the speed controller and determine the need to select a two-fold integrating speed control system.

The choice of this speed control method determines the need to select the shape of the desired logarithmic amplitude-frequency characteristic (LAFC), as shown in Fig. 6 [7]. The LAFC (Fig. 6) can be used to write the desired transfer function of the open speed loop in the following form:

$$W_{des}(p) = \frac{k_\Sigma(1 + T_2p)}{p^2(1 + T_3p)},$$

(11)

where k_Σ – desired transmission coefficient of the open speed loop; T_2, T_3 – time constants inversely proportional to the coupling frequencies of the mid-frequency part of the LAFC circuit, placed relative to the cut-off frequency Ω_{sl}, respectively in the low-frequency and high-frequency area of the LAFC.

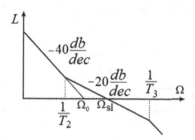

Fig. 6. Desired LAFC of the open speed loop.

The peculiarity of the speed controller synthesis is that the desired LAFC is selected (Fig. 6) and the transfer function (11) is recorded. These features require a more detailed description of the procedure for the synthesis of the speed controller (SC):

1. Set by the cut-off frequency Ω_{sl}, by an order of magnitude lower the ratio $1/T_{el}$. This choice of Ω_{sl} makes it possible to neglect the small time constant T_{el}.
2. The transfer function of the speed controller is recorded, the structure of which is easily determined from the product of sequentially connected transfer functions – the controlled object and the speed controller, which allows obtaining the Eq. (11). Hence, the desired transfer function of the SR is as follows:

$$W_{sc}(p) = \frac{k_{sc}(1 + T_2 p)}{p(1 + T_3 p)}. \tag{12}$$

3. If the conditions of items 1, 2 are satisfied, then following the recommendations of [7, 11], the parameters of the speed controller can be determined based on the following ratios:

$$T_2 = \frac{M^2}{\Omega_{sl}(M - 1)^2}; \quad \Omega_0 = \sqrt{\frac{\Omega_{sl}}{T_2}}; \quad k_{sc} = \frac{\Omega_0^2 J_\Sigma}{k_t}; \quad T_3 = \frac{1}{\Omega_0}\sqrt{\frac{M(M - 1)}{(M + 1)^2}}, \tag{13}$$

where M = 1.1 – index of oscillation.

The synthesis of the position regulator is performed according to a typical method adopted in the theory of the electric drive [7], taking into account the fact that the main function of the electric drive of the hydraulic distributor should be aimed at the accurate processing of the proportional movements of the working element. Since the physical processes in the position control circuit are slow, it is possible to select the position controller in the form of a proportional block with the transmission coefficient k_{pr}. Then the structural diagram of the position contour will similar to the one shown in Fig. 7.

In accordance with Fig. 7, the transfer function of the controlled object is written as follows:

$$W_{pco}(p) = \frac{k_s k_p}{[1 + (T_{\mu s} + T_{enc})p]p},$$ (14)

where $T_{\mu c} = 1/\Omega_{sl}$ – uncompensated constant of the speed loop; T_{ps} – time constant of the position sensor.

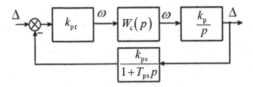

Fig. 7. Structural diagram of the position contour of the position electric drive.

Following the recommendations from [7], let us adjust the position regulator to the technical optimum. In this case, the desired position contour transfer function must correspond to the expression:

$$W_{des.p}(p) = \frac{1/k_{enc}}{2T_{\mu p}p(1 + T_{\mu p}p)},$$ (15)

where $T_{\mu p}$ – uncompensated constant of the position contour; k_{ens} – transmission factor of the position sensor.

Dividing (15) by (14) we get the desired transfer function of the position regulator in the form:

$$W_{c.p}(p) = \frac{1}{2 \cdot T_{\mu p} \cdot k_s \cdot k_p \cdot k_{enc}}.$$ (16)

3.1 Control Logic

The dynamic properties of the synthesized position control system can be determined only on the basis of experimental studies directly at the control object. The functional diagram of the electric drive of the high-pressure press hydraulic distributor is made according to the diagram shown in Fig. 3.

The power part of the position electric drive designed for the 30,000-ton press has the following parameters:

- induction motor: $P_n = 11$ kW, $U_n = 220/380$ V, $I_n = 36/21$ A, $n_n = 2,950$ rpm, $\eta = 0.88$, $\cos\varphi_{1n} = 0.9$;
- tachometer generator: $P_n = 20$ W, $k_{tg} = 57.5$ mV/rpm; reduction gear: $i = 51$, $J_{\Sigma} = 0.0288$ kg * m^2.
- position sensor: $k_{ps} = 19.1$, $T_{ps} = 0.01$ s.

There is a need for experimental studies both at no-load and under load to fully evaluate the dynamic properties of the electric drive.

3.2 Results

The dynamic properties of the synthesized position control system can be determined only on the basis of experimental studies directly at the control object.

Figure 8 shows oscillograms characterizing the dynamic properties of the electric drive speed control loop, which reflect the processes occurring during the idle reverse of the electric drive at maximum speeds.

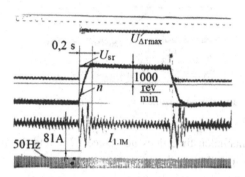

Fig. 8. Oscillogram characterizing dynamic properties of the speed control loop.

Figure 9 shows oscillograms characteristic of the most severe operating mode of the electric drive, which occur when the hydraulic valve is opened.

The oscillograms in Fig. 10 show the behavior of the electric drive, when during the operation of the hydraulic press there are modes with peak load rise occurring in the hydraulic distributor.

3.3 Discussion

As shown by the oscillograms in Fig. 8, the reverse of the electric drive occurs in a time of 0.2 s, while the speed overcontrol does not exceed 5%, which indicates the high dynamic properties of the system, which were laid down by the selection of the LAFC system in Fig. 6. As shown in Fig. 9, the maximum load torque occurs when the valve is fully closed, and the load gradually decreases as the valve opens.

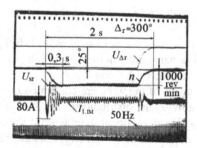

Fig. 9. Adjustment of the maximum mismatch angle by electric drive.

As shown by the results of the experiment (Fig. 10), transient processes in working cycles with peak load proceed smoothly, almost without readjustment and, at the same time, at high speed.

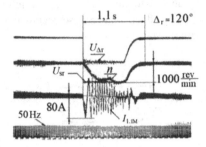

Fig. 10. Adjustment of mismatch at peak load.

The maximum mismatch time does not exceed 2 s, and the positioning error, as measured on the unit, is not more than 2°.

4   Conclusion

The performed theoretical and experimental studies of the energy indicators of the DFC-IM system show that it is advisable to complete the electric drive of the ultra-high pressure hydraulic press with only symmetrical circuits: six-pulse or three-pulse thyristor transducers for three-phase induction motors.

The use of a microprocessor generator of sinusoidal signals, which generates signals for setting the current by the phases of an induction motor, should have an algorithm in which the known relationships between the amplitude, phase and frequency of the IM stator field are maintained.

The introduction of a positive feedback torque to the regulator input, as well as its critical adjustment, allowed the torque contour to be optimized and presented as an aperiodic link. At the same time, the choice of the shape for the LAFC as shown in Fig. 2 made it possible to adjust the speed controller to the two-fold integrating.

Industrial tests of the complete position electric drive of the hydraulic distributor of the 30,000-ton press within the DFC-IM system confirm that the adoption as a criterion of satisfactory transient processes $M = 1.1$ and the choice of $\Omega_{sl} = 25$ 1/s made it possible to obtain the parameters of the electric drive ECS: $\Omega_0 = 10.7$ 1/s; $T_2 = 0.44$ s; $k_\Sigma = 173$; $T_3 = 0.008$ s; $k_{sc} = 0.47$; $k_c = 12.6$; $k_p = 0.0196$; $T_{\mu p} = 0.03$ s; $k_{ps} = 2.12$. The implementation of these parameters in the system fully meets the requirements for electric drives of such mechanisms.

The main theoretical results of the work are confirmed by a general experiment on the industrial model of the hydraulic distributor of the 30,000-ton press. The practical recommendations for choosing the rational diagrams for the layout of power circuits of the DFC-IM system can be used by design engineers engaged in automation when creating AC electric drives for positioning mechanisms.

References

1. Ovchinnikov, A.: Progressive Technological Processes of Cold Forming (in Russ.) (Progressivnye tekhnologicheskie processy holodnoj shtampovki). Mashinostroenie, Moscow (2001)
2. Kadyrov, I., Polyaninov, G., Matekova, G.: Effective control of hydraulic presses with electromechanical distributors (in Russ.) (Effektivnoe upravlenie gidravlicheskimi pressami s elektromekhanicheskimi raspredelitelyami). Problemy Avtomatiki i Upravleniya **1**, 233–236 (2010). http://pau.imash.kg/index.php/pau/article/view/121. Accessed 10 Oct 2023
3. Vasile, I., Tudor, E., Sburlan, I.-C., Matache, M.-G., Cristea, M.: Optimization of the electronic control unit of electric-powered agricultural vehicles. World Electr. Veh. J. **14**, 267 (2023). https://doi.org/10.3390/wevj14100267
4. Liao, W.-H., Wang, S.-C., Liu, Y.-H.: Learning switched mode power supply design using MATLAB/SIMULINK. In: Proceedings of TENCON 2009 - 2009 IEEE Region 10 Conference, 23–26 January 2009 (2009). https://doi.org/10.1109/TENCON.2009.5395993
5. Genliang, L., Xinjin, W., Jun, Y., Guoliang, Z., Yunbing, W.: The modeling and analysis of asynchronous motor based on matlab/simulink. In: Proceedings of 2010 International Conference on Computing, Control and Industrial Engineering, Wuhan, China, pp. 442–445 (2010). https://doi.org/10.1109/CCIE.2010.117
6. Kadyrov, I., Karaeva, N., Andarbekov, Z., Kadyrkulova, K.: Features of designing a variable-frequency electric drive control system with a microprocessor-based sinusoidal signal generator. Commun. Comput. Inf. Sci. **1304**, 203–220 (2021)
7. De Doncker, R.W., Pulle, D.W.J., Veltman, A.: Advanced Electrical Drives: Analysis, Modeling, Control. Springer, Cham (2020). https://doi.org/10.1007/978-3-030-48977-9
8. Bochkarev, I., Kadyrov, I.: Microprocessor control device according to DFC-IM system of the electric drive of an excavator. Izvestiya Vysshikh Uchebnykh Zavedenii. Elektromekhanika **5**, 25–30 (2007)
9. Kadyrov, I., Postnov, A.: Development of a mathematical model of a sinusoidal signal generator (in Russ.) (Razrabotka matematicheskoj modeli generatora sinusoidal'nogo signala). Izvestiya KGTU im. Razzakova **32**(1), 153–157 (2014). (in Russian)
10. Ushkov, A., Kolganov, A.: Analysis and modeling of AC electric drive with PFC. In: Proceedings of 2016 IX International Conference on Power Drives Systems (ICPDS), Perm, Russia, pp. 1–5 (2016). https://doi.org/10.1109/ICPDS.2016.7756726
11. Ramshaw, R.: Power Electronics: Thyristor Controlled Power for Electric Motors. Springer, Dordrecht (2012). https://doi.org/10.1007/978-94-011-6916-5

Development of a Mathematical Model to Study the Energy Indicators of Electric Drives Using the DFC-IM System

Ishembek Kadyrov[1]([⊠]) [iD], Nurzat Karaeva[1], and Alymbek uulu Chyngyzbek[2]

[1] Skryabin Kyrgyz National Agrarian University, Mederov Str. 68, 720005 Bishkek, Kyrgyz Republic
bgtu_kg@mail.ru

[2] Razzakov Kyrgyz State Technical University, Ch. Aitmatov Ave. 66, 720044 Bishkek, Kyrgyz Republic

Abstract. The paper is devoted to the peculiarities of constructing a mathematical model of a frequency-controlled electric drive, and with regard to powerful units justifies the need to power the windings of an induction motor with a short-circuited rotor from the output of a direct frequency converter (DFC) assembled according to a symmetrical scheme using six-pulse or three-pulse thyristor converters. The justifications are given using the example of electric drives of ditching machines and hydraulic presses. These machines perform operations that are fundamentally different from each other, but have the same structure of the electric drive control system. At the same time, these differences do not affect the choice of the basic structure for creating a mathematical model of a frequency-controlled electric drive. The givenblock diagram, which is common for both machines and is made taking into account the power of the convertible electric energy, makes it possible to justify the issue of powering the stator windings of the induction motor from the output of the direct frequency converter, if the converter operates in the current source mode. The chosen processing units involved in various technical processes give the basis for the possibility of applying the results of the study to any units where the AC electric drives are used according to the DFC-IM system. The developed mathematical model will allow studying the energy characteristics of the consumed electric energy during the operation of the electric drive in a steady mode. Equations describing the power part of the DFC-IMsystem are compiled using the state variable method. At the same time, such elements of the power circuit as the transformer and the induction motor are represented by their replacement schemes with reduced parameters; thyristors of the frequency converter are replaced – in a closed state by their dynamic resistances R_T, in a closed state – by zero current sources. The obtained mathematical model of the electric drive according to the DFC-IMsystem allows studying the energy processes occurring during energy conversion and formulating practical recommendations for choosing the rational design schemes of power circuits.

Keywords: Mathematical Model · Direct Frequency Converter · Induction Motor · Thyristor Converter · Block Diagram · Software · Spectral Analysis

V. Jordan et al. (Eds.): HPCST 2023, CCIS 1986, pp. 234–247, 2024.
https://doi.org/10.1007/978-3-031-51057-1_18

1 Introduction

Various technological units are used in machine-building production, thus making it possible to achieve high accuracy of product manufacturing, which is one of the main indicators of the development of production forces. The accuracy requirements for modern machines and units can be satisfied if the permissible deviations do not exceed the tolerance limits calculated in 0.002 mm.

Such indicators of product manufacturing accuracy in technological units are implemented in the presence of adaptive electric drive control systems of the main mechanisms of the equipment involved in the manufacture of finished products. This approach in the control of such units indicates the achieved conditions for optimizing the flow of technological processes, i.e., accuracy and productivity control in production. The studies by many authors aimed at improving the accuracy of the geometric dimensions of products led to the consensus that this indicator is used not only for the direct manufacture of products for their intended service, but also is one of the prerequisites for their long-term operation without losing the initial accuracy. Therefore, this may refer to a certain "margin of accuracy", which can be achieved only if resources are created in the products to compensate for the physical wear during their operation [1].

The design of AC electric drives is currently a priority task. Moreover, in the process of design, the authors comprehensively studying the designed object in laboratory conditions resort to creating a physical as well as mathematical model. The purpose of this paper is to develop a mathematical model of a frequency-controlled electric drive to power the winding of an induction motor (IM) from a direct frequency converter (DFC).

2 Principles ForPositioning Accuracy

The mathematical model requires objects, which dynamic and static properties of the electric drive within the DFC-IM system are sufficiently studied. Such electric drives, where the authors were directly involved in the development of control systems, include the AC electric drives of the main mechanisms of the walking excavator [2, 3], as well as the frequency-controlled electric drive of the electromechanical distributor of the ultra-high pressure hydraulic press [4].

It should be noted that these objects differ not only in their design, but also in the modes of operation of electric drives when performing technological operations. Electric drives of the main mechanisms of the walking excavator operate in a repeated short-term mode with frequent reverses when the load changes within large limits, including when the working element is locked. The electric drive control systems of the main mechanisms of the ditching machine solve the problem of ensuring reliable operation by preventing breakdowns in mechanical links. The electric drive of the hydraulic distributor of the ultra-high pressure hydraulic press operates in a long-term mode, and the reverse of the electric drive is only needed upon the final adjustment of positioning or when the slide is removed from the product. The control system of this object solves the tasks of product manufacturing accuracy indicated in the introduction.

The analysis of technical requirements for these objects confirmed that the best indicators satisfying the reliable control are systems in which stator windings of an induction motor are connected to the output of a direct frequency converter assembled from the required number of reversible DC transducers. Besides, it was theoretically and experimentally proved that the tasks of frequency control of the torque and speed of an asynchronous electric drive can be solved only if the flux linkage vector of the rotor Ψ_2 is maintained at a constant level Ψ_{2max}, and the constancy of the rotor flux linkage in these systems can be ensured indirectly – through compensation [5].

This control method ensures the relationship between the engine torque and the absolute slip, which can be presented as follows:

$$(1 + T_e p)T = \beta \cdot s_a \cdot \omega_{0.n}. \tag{1}$$

Electromagnetic inertia in the motor electromagnet torque control system in Eq. (1) is taken into account by introducing a small electro-magnetic time constant T_e.

In order to satisfy the control conditions described by Eq. (1), a compensation method is used by introducing a positive coupling according to speed ω achieved through its direct measurement, which is subsequently used to generate a stator current frequency based on the following ratio:

$$\omega_{0.el} = \omega p_p + \omega_{0.el.n} \cdot s_a. \tag{2}$$

The method of electric drive speed control according to Eq. (2) is achieved due to the introduction of positive speed feedback to the input of the torque regulator thus ensuring the astatic control of the torque and speed during its critical adjustment within constraints limited by the voltage margin of the converter.

To achieve the purpose of this paper, let us take the basic diagram shown in Fig. 1 to create a mathematical mode of a frequency-controlled electric drive within the DFC-IM system, which is convenient when designing software for a mathematical model of an electric drive in the DFC-IM system.

The block scheme in Fig. 1 uses PI– a link that provides astatic speed control as a speed controller (SC). The analysis of mechanical characteristics makes it possible to conclude that the engine torque limitation is provided in the system in accordance with the input signal of the speed controller U_{tr}, as is the case in similar systems of subordinate control of the torque and speed in DC electric drives [5]. The microprocessor-based sinusoidal signal generator specified in Fig. 1 is designed to generate currents in IM phases with adjustable frequency, amplitude and phase [6, 7].

The studies of the dynamic processes of electric drives of identified mechanisms showed the possibility of using a frequency-controlled electric drive within the DFC-IM system according to the structure in Fig. 1 mainly in high-power mechanisms. However, the assessment of energy characteristics for the class of mechanisms in which the electric drive is used according to the DFC-IM system can only be performed by comparing the corresponding mathematical dependencies.

The developed mathematical model will allow studying the energy characteristics of the consumed electric energy during the operation of the electric drive in a steady mode. The analysis of the shape of the currents of the windings of the stator of the engine, transformer and other components of electrical equipment will allow assessing the components of the energy consumption from the network.

For the clarity of the developed mathematical model of the electric drive let us assumes that the induction motor receives power from the output of a direct frequency converter assembled according to a symmetrical three-pulse scheme. The modeling results will be compared by oscillograms of the phase currents of the induction motor on a laboratory bench, where the direct frequency converter is assembled according to a six-pulse scheme.

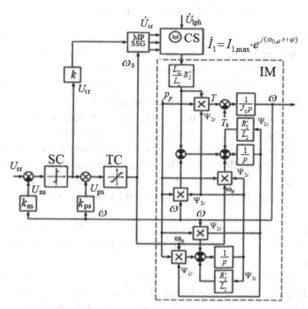

Fig. 1. Block diagram of the control of hydraulic directional valve electric drive using the DFC-IM system: CS– current source supplying the IM winding; IM – two-phase model of the generalized machine when powered by the CS; SC– PI-speed controller; TC– proportional torque controller; MPSSG– three-phase microprocessor-based sinusoidal signal generator.

2.1 Control Panel Functions

The study of the energy characteristics of the electric drive using the DFC-IM system implies that the mathematical model of the power part is presented by a variable structure, while it is rational to use the known digital model with additional algorithms for processing the energy parameters.

In this case, the possibilities of analyzing the results of theoretical studies in a system with a variable structure of the power part, supplemented by various information processing methods, are provided by the construction of the digital model itself. The most significant features characterizing this model in terms of applicability in solving the tasks set in this study are as follows:

1. Structure blocking effect.
2. Output units providing the following:

a. possibility of setting arbitrary initial conditions of the studied processes (torque, flux linkage, phase currents, etc.);
b. possibility of printing the required values with several scales at arbitrary points;
c. processing of any variables as required;
d. introduction of the structure and parameters of the system by describing each section in the theory of electrical circuits.

2.2 Generation of Mathematical Model Equations

When describing the DFC-IM system, the equations of an induction motor with a short-circuit rotor are presented in a three-phase coordinate system non-moving relative to the stator windings. The choice of such a coordinate system allows considering electromagnetic processes in the stator circuit in non-moving axes relative to it, taking into account all three degrees of freedom of a three-phase system. The need for such a description is caused by the asymmetry of voltages in the stator windings of the motor, due to the presence of thyristor converters, even when the power circuit of the system is arranged with symmetric schemes. At the same time, the short-circuit rotor of the induction motor has no asymmetry, which allows describing electromagnetic processes in it in a coordinate system rotating at an arbitrary speed. This approach, when generating the rotor circuit equations, allows using constant values of mutual induction between stator and rotor windings [8].

It is necessary to emphasize the peculiarities of compiling equations of the IM electromechanical characteristic both for the IM control flow chart and for building the IM mathematical model.

The rearrangement of the equations of the generalized machine shown in Fig. 1 was performed in axes $0, x, y$, which made it possible to represent the variables of the two-phase model as rotating at the synchronous speed of the machine.

The peculiarities of the IM mathematical model, which receives power from the output of the direct frequency converter, include the selection of axes $0, \alpha, \beta$, which allows, when compiling equations, considering IM voltages and currents in the form of variables with the same frequency.

In a matrix form these equations will be written as follows:

$$\frac{[U_{1\alpha\beta}]}{[U_{2\alpha\beta}]} = \frac{[R_{1\alpha\beta} + pL_{1\alpha\beta}][pL_{12\alpha\beta}]}{[pL_{12\alpha\beta} + k_{\alpha\beta}L_{12}][R_{2\alpha\beta} + pL_{2\alpha\beta} + k_{\alpha\beta}L_2]} \cdot \frac{[i_{1\alpha\beta}]}{[i_{2\alpha\beta}]};$$

$$U_{1\alpha\beta} = \begin{vmatrix} U_{10} \\ U_{1\alpha} \\ U_{1\beta} \end{vmatrix}; \quad i_{1\alpha\beta} = \begin{vmatrix} i_{10} \\ i_{1\alpha} \\ i_{1\beta} \end{vmatrix}; \quad U_{1\alpha\beta} = \begin{vmatrix} U_{20} \\ U_{2\alpha} \\ U_{2\beta} \end{vmatrix}; \quad i_{2\alpha\beta} = \begin{vmatrix} i_{20} \\ i_{2\alpha} \\ i_{2\beta} \end{vmatrix}; \tag{3}$$

where L_1, R_1 – total inductive and active resistance of the stator phase of a two-phase machine; L_2, R_2 – total inductance and active resistance of the rotor phase of a two-phase machine reduced to the stator; L_{12} – mutual inductance of a rotor and stator windings of a two-phase machine at coincidence of their axes; p_p – number of pole pairs; i, U – current, voltage; index 1 – stator values, 2 – rotor values; p – differentiation operator.

Resistance and inductance of stator and rotor windings in a matrix form are represented as follows:

$$R_{1\alpha\beta} = \begin{vmatrix} R_1 & 0 & 0 \\ 0 & R_1 & 0 \\ 0 & 0 & R_1 \end{vmatrix}; \quad R_{2\alpha\beta} = \begin{vmatrix} R_2 & 0 & 0 \\ 0 & R_2 & 0 \\ 0 & 0 & R_2 \end{vmatrix}; \quad k_{\alpha\beta} = \begin{vmatrix} 0 & 0 & 0 \\ 0 & 0 & p\omega \\ 0 & -p\omega & 0 \end{vmatrix};$$

$$L_{1\alpha\beta} = \begin{vmatrix} L_1 - L_{12} & 0 & 0 \\ 0 & L_1 & 0 \\ 0 & 0 & L_1 \end{vmatrix}; \quad L_{2\alpha\beta} = \begin{vmatrix} L_2 - L_{12} & 0 & 0 \\ 0 & L_2 & 0 \\ 0 & 0 & L_2 \end{vmatrix}; \quad L_{12\alpha\beta} = \begin{vmatrix} L_{12} & 0 & 0 \\ 0 & L_{12} & 0 \\ 0 & 0 & L_{12} \end{vmatrix}. \tag{4}$$

Using real currents and voltages of IM phases in three-phase axes as variables, we get the voltage equilibrium equations:

$$\frac{[U_{1abc}]}{[U_{2abc}]} = \frac{\left[R_{1abc} + pL_{1abc}\right]\left[pL_{12abc}\right]}{\left[pL_{12abc} + k_{abc}L_{12}\right]\left[R_{2abc} + pL_{2abc} + k_{abc}L_2\right]} \cdot \frac{[i_{1abc}]}{[i_{2abc}]}. \tag{5}$$

To ensure invariance of power during the transition from two-phase to three-phase coordinate system, the matrices of equation parameters must be connected by linear transformation [9]:

$$Z' = T \cdot Z \cdot T^{-1},$$

where Z' – matrix of parameters in a 3-phase system; Z– matrix of parameters in a 2-phase system.

$$T = \sqrt{\frac{2}{3}} \begin{vmatrix} \frac{\sqrt{2}}{2} & 1 & 0 \\ \frac{\sqrt{2}}{2} & -\frac{1}{2} & \frac{\sqrt{3}}{2} \\ \frac{\sqrt{2}}{2} & -\frac{1}{2} & -\frac{\sqrt{3}}{2} \end{vmatrix}. \tag{6}$$

The interaction between the mechanical and electrical parts of the IM is ensured by means of the electromotive force in the windings as a result of the mechanical movement of the machine rotor. In matrix form, this interference is expressed as follows:

$$E_{2abc} = \begin{vmatrix} E_{2a} \\ E_{2b} \\ E_{2c} \end{vmatrix} = \frac{p_p\omega}{\sqrt{3}} \begin{vmatrix} 0 & 1 & -1 \\ -1 & 0 & 1 \\ 1 & -1 & 0 \end{vmatrix} \cdot \left(\begin{vmatrix} i_{1a} \\ i_{1b} \\ i_{1c} \end{vmatrix} L_{12} + \begin{vmatrix} i_{2a} \\ i_{2b} \\ i_{2c} \end{vmatrix} L_2 \right). \tag{7}$$

Taking into account that the IM rotor is short-circuited ($U_{2a} = U_{2b} = U_{2c} = 0$) and considering the influence of the electromotive force, the Eq. (5) for a three-phase coordinate system can be written as follows:

$$\frac{[U_{1abc}]}{[-E_{2abc}]} = \frac{\left[R_{1abc} + pL_{1abc}\right]\left[pL_{12abc}\right]}{\left[pL_{12abc}\right]\left[R_{2abc} + pL_{2abc}\right]} \cdot \frac{[i_{1abc}]}{[i_{2abc}]}. \tag{8}$$

The obtained equations of electrical equilibrium (8) correspond to the substitution scheme given in Fig. 2, which is valid for arbitrary operating modes of the machine. In this scheme, the presence of flux linkages between the stator and rotor windings of the IM is assumed, but not conditionally shown.

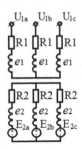

Fig. 2. IM replacement scheme.

As a result of the interaction of currents flowing through the IM windings, an electromagnetic torque develops, which is determined using the known ratios [5, 10].

$$T = \frac{p_p L_{12}}{\sqrt{3}}[i_{1a}(i_{2c} - i_{2b}) + i_{1b}(i_{2a} - i_{2c}) + i_{1c}(i_{2b} - i_{2c})]. \tag{9}$$

The mechanical part of the IM is described by the motion equation, provided that there is a rigid reduced mechanical connection:

$$T - T_s = J_\Sigma \frac{d\omega}{dt}, \tag{10}$$

where J_Σ – total inertia moment of the rotor; T_s – total resistance moment of the rotor.

Equations (8–10) fully describe the processes of electromechanical energy conversion in an induction three-phase engine with a short-circuited rotor.

Equations describing the power part of the DFC-IM system are compiled using the state variable method. In this case, the transformer and induction motor supplying power are represented by their replacement schemes with the given parameters; thyristors of the frequency converter are replaced: in the open state – by their dynamic resistances R_T, in the closed state – by zero current sources.

These equations allow determining currents and voltages on any branches of the circuit at an arbitrary moment in time and, together with the IM equations, describing the electromechanical processes of the entire DFC-IM system.

The equations are generated automatically in the program built on the basis of the state variable method, and the user prepares information about the topology and parameters of the required power scheme. The user numbers all nodes and branches of the replacement circuit in a strictly defined sequence, and then enters information into the PC. The procedure for numbering the nodes and branches of the circuit, as well as the sequence of information entry, is connected with the need to ensure the correct operation of programs that simulate the functioning of pulse-phase control circuits (PPCC) and monitor the state of thyristors.

3 Block Diagram of Calculation Software in the DFC-IM System

An outline flowchart of the software used to solve the assigned tasks is shown in Fig. 3. The main software elements and their interconnection are shown in Fig. 4.

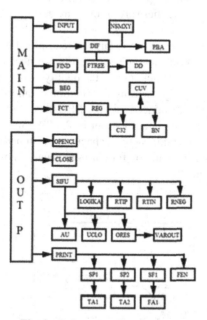

Fig. 3. Software outline flowchart.

The INPUT subprogram enters the system power diagram data from the external memory. According to the data describing the scheme topology, the matrices necessary to obtain the equations of the power circuit are built in the subprograms DIF, FTREE, DD, PBA. The resulting matrix expressions are solved in subprograms: FCT – equations of state, VAROUT and UCLO – equations of voltages and currents. With each change in the state of thyristors, the process of forming matrices is automatically repeated. The most complex automatic generation of equations of the power scheme of the DFC-IM system is carried out in the considered software modules.

At the same time, the access is provided at the necessary points of the program to the subprograms compiled by the user and describing the system of equations of the drive and the required type of data processing that sets the mode of the experiment on the model.

So, the initial conditions are set, the algorithm for reaching the working point, and the parameters of the control system are determined in the BEG subprogram. The REG subprogram, which is accessed from the subroutine of the right-hand solution of differential equations, is designed for the right-hand parts of these equations in the standard Cauchy form. In the UREG module, the signals coming to the input of the pulse-phase control system (PPCC) of the DFC are summed up.

The SIFU contains a detailed description of the operation of the PPCC of one monoblock reversing converter, which during the operation of the program is repeated in a cycle as many times as rectifiers in the DFC, with the corresponding replacement of the numbers of the starting valves. The type of PPCC adjustment characteristic is specified by the user in the AI subprogram. The algorithms of a line operation section for separate control of reversing rectifier valve groups are described by the user in LOGIKA subprogram.

The ease of the program operation is largely determined by the form of the information provided and the ability to process it using the necessary methods. Thus, in order to obtain the integral characteristics of the quality of the consumed energy by the electric drive, an algorithm for processing information by an integral method for determining energy indicators is introduced into the REG subprogram, and final processing and printing of the final results is carried out in the TAI and TA2 program modules, where information on the full and reactive capacities consumed by the DFC-IM system is received. FA1 processes information for harmonic analysis of the output current of a direct frequency converter.

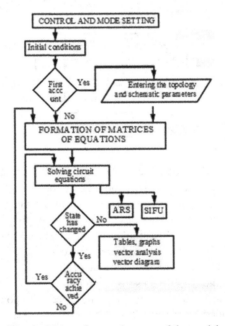

Fig. 4. Main software elements of the model.

3.1 Control Logic

To assess the conformity of the obtained results of the model with the energy and power indicators of the real system, as well as for the complete analysis of the power circuits of the DFC-IM system, we designed a laboratory stand, which, as a result of experimental studies, made it possible to confirm the adequacy of the obtained results on a mathematical model. The functional diagram of the laboratory unit shown in Fig. 5 uses the maximum number of thyristors in the DFC units.

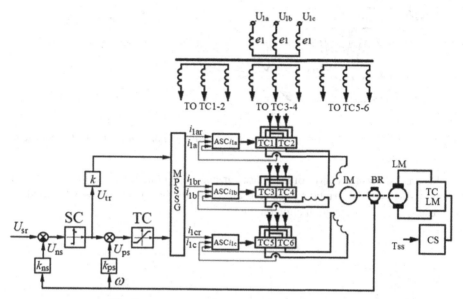

Fig. 5. Functional diagram of the laboratory bench.

An electric drive within the TP-D system operating in the mode of an adjustable torque source serves the load device. The power part of the DFC-IM system consists of the DFC equipped with thyristor converters assembled according to the bridge joint in the amount of six power units. Each reversible thyristor converter of the DFC is included in a combined control system to give it the properties of a current source.

A A5I-4A engine with rated parameters is used as an operating motor: $P_n = 4.5$ kW; $n_n = 1{,}440$ rpm; $U_n = 220/380$ V; $I_n = 17/9.9$ A; $\eta_n = 83.5\%$; $\cos\varphi_n = 0.83$; $R_1 = 1.5$ Ω; $X_{sc} = 7.4$ Ω.

3.2 Results

The experimental studies to identify the adequacy of the mathematical model presented by the block diagram in Fig. 3 and the physical model with the real configuration of the electric drive system shown in Fig. 5 were conducted by reading the oscillograms of phase currents and induction motor voltages in the closed ACS.

In this case, the mathematical model of the induction electric drive built according to the scheme in Fig. 4 and the electrical circuit of replacement are presented in the form of units and branches, numbered in a sequence that ensures the direct operation of programs that simulate the control system of thyristor monoblocks.

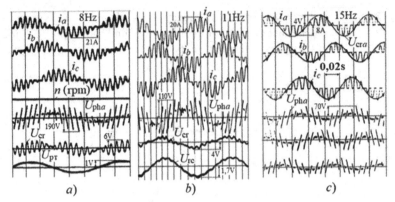

Fig. 6. Oscillograms of IM currents and voltages in the electric drive using the DFC-IM system.

Figure 6 shows the curves of IM phase currents and voltages during the operation of the DFC-IM system in static modes, and Fig. 7 shows the curves of currents based on the results of the system simulation on the PC.

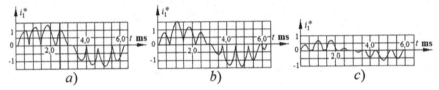

Fig. 7. Phase currents of a three-phase IM with symmetrical three-pulse DFC: (a) $T_s = T_n$, (b) $T_s = -T_n$, (c) $T_s = 0$.

3.3 Discussion

Let us consider the general patterns of operation of a three-pulse DFC. To do this, let us analyze the shape of the curves of the output voltage and the corresponding input currents of the DFC by the switching functions and the introduction of generally accepted assumptions: all elements of the system are ideal, the switching of the valves is instantaneous, the output currents of the converter are continuous and sinusoidal, the DFC operates on a symmetrical n-phase R-L load [9, 11].

The calculated dependencies of the reactive power consumption of the ideal system are shown in Fig. 8. These characteristics show that the minimum consumption of reactive energy Q corresponds to the idle mode of the electric drive. As the engine load increases, and hence the output current of the DFC increases, the consumption level Q^* increases, and is faster in the generator mode than in the motor mode. The level Q^* also increases with a decrease in the output frequency of the DFC, which is explained by a decrease in the relative output voltage of the DFC [11].

To clarify the obtained estimates of energy characteristics, let us use a mathematical model of an asynchronous electric drive. The power part of the symmetrical DFC with a three-phase IM is represented by an electrical replacement scheme, where nodes and branches must be numbered in a sequence ensuring the correct operation of programs that simulate the control system of thyristor monoblocks. Having completed this step and ensuring the settings of the real electric drive system in the model, let us study the quality of the formation of phase currents of the engine (DFC output currents) in the closed ECS.

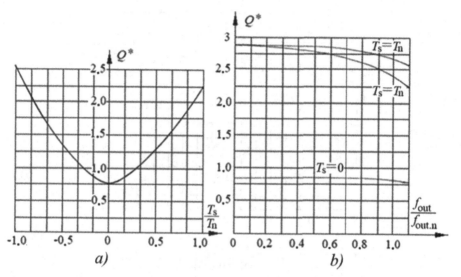

Fig. 8. Consumption dependencies Q^* of the ideal DFC on load (a) and output frequency (b)

Since the stator currents are formed in a closed motor speed control system that constantly affects the amplitude, frequency and phase of the current setting signal, and therefore the current shape itself, it is desirable to have averaged estimates of the spectral composition of the obtained curves. The histograms shown in Fig. 9 reflect the spectral composition of all DFC output currents over three periods.

The harmonic analysis shows that at the rated load of the electric drive in motor and generator modes, the main generated current harmonic is dominant. The distortion factor k_{i1}, which determines the content of the fundamental harmonic in the motor current curve, equals 0.933 for motor and 0.921 for generator modes.

As expected, the distorting components of the IM phase currents are represented by a wide spectrum, however, the greatest distortions are introduced by the components characteristically expressed in the output voltage curve of the ideal DFC. Components with relatively low frequencies have a significant influence on the shape of currents, while high-frequency components that distort the shape of the current are filtered.

At the same time, the amplitudes of the distorting components in the entire range of changes in the load of the electric drive compared to the fundamental harmonic change slightly, which explains the low current distortion coefficient (0.73) during idle operation of the engine. Besides, additional losses in copper from distorting components depend little on the load. Indeed, when the electric drive is operating without load, additional losses make 8%, and increase at the nominal load to 12–15% relative to nominal losses.

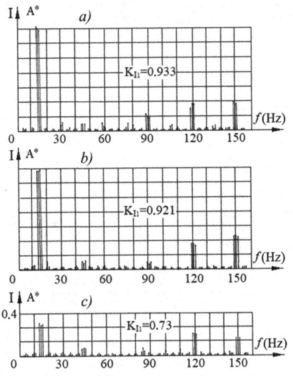

Fig. 9. Harmonic composition of currents in a three-phase IM with symmetrical DFC: (a) $T_s = T_n$, (b) $T_s = -T_n$, (c) $T_s = 0$.

The considered power circuit makes it possible to form a symmetrical system of currents in IM phases. Thus, the asymmetry of the amplitudes of the main harmonics of IM phase currents in steady-state operating modes at the rated load of the electric drive does not exceed 5%.

4 Conclusion

The oscillograms shown in Fig. 6, c indicate that the smooth component of the IM phase currents and the current setting voltages generated by the micro-processor sinusoidal signal generator coincide in phase, which is very important for the stable operation of the electric drive.

The performed theoretical and experimental studies of energy-static indicators of the DFC-IM system show that it is advisable to complete powerful electric drives of production units only with symmetric schemes of a three-phase IM when powered from a six-pulse DFC.

The main theoretical results of the work were experimentally confirmed on a laboratory unit and an industrial model of ditching machines and a hydraulic distributor of a 30,000-ton press.

References

1. Suslov, A.: Scientific Foundations of Mechanical Engineering Technology (in Russ.) (Nauchnye osnovy tekhnologii mashinostroeniya). Mashinostroenie, Moscow (2002)
2. Bochkarev, I., Kadyrov, I.: Microprocessor control device according to DFC-IM system of the electric drive of an excavator. Izvestiya Vysshikh Uchebnykh Zavedenii. Elektromekhanika 5, 25–30 (2007)
3. Bochkarev, I., Kadyrov, I.: Optimization of parameters of control device for electromechanical system of excavators. Russ. Electr. Eng. 80(2), 61–66 (2009)
4. Kadyrov, I., Karaeva, N., uulu Azamat, B.: Principles of building remote control by DFC-ID system for press hydraulic distributor with pressure of 30,000 tons (in Russ.) (Principy postroeniya distancionnogo upravleniya elektroprivodom po sisteme NPCh-AD dlya gidro-raspredelitelya pressy s davleniem v 30000 tonn). Izvestiya KGTU im. Razzakova 51(3), 95–106 (2019)
5. De Doncker, R.W., Pulle, D.W.J., Veltman, A.: Advanced Electrical Drives: Analysis, Modeling, Control. Springer Cham (2020). https://doi.org/10.1007/978-3-030-48977-9
6. Kadyrov, I, Postnov, A.: Development of a mathematical model of a sinusoidal signal generator (in Russ.) (Razrabotka matematicheskoj modeli generatora sinusoidal'nogo signala). Izvestiya KGTU im. Razzakova 32(1), 153–157 (2014)
7. Kadyrov, I., Karaeva, N., Andarbekov, Z., Kadyrkulova, K.: Features of designing a variable-frequency electric drive control system with a microprocessor-based sinusoidal signal generator. Commun. Comput. Inf. Sci. 1304, 203–220 (2021)
8. Diaz, A., Saltares, R., Rodriguez, C., Nunez, R.F., Ortiz-Rivera, E.I., Gonzalez-Llorente, J.: Induction motor equivalent circuit for dynamic simulation. In: Proceedings of 2009 IEEE International Electric Machines and Drives Conference, Miami, FL, USA, pp. 858–863 (2009). https://doi.org/10.1109/IEMDC.2009.5075304
9. Liao, W.-H., Wang, S.-C., Liu, Y.-H.: Learning switched mode power supply design using MATLAB/SIMULINK. In: Proceedings of TENCON 2009 - 2009 IEEE Region 10 Conference, 23–26 January 2009 (2009). https://doi.org/10.1109/TENCON.2009.5395993
10. Kodkin, V., Anikin, A.: On the physical nature of frequency control problems of induction motor drives. Energies 14(14), 4246 (2021). https://doi.org/10.3390/en14144246
11. Dinolova, P., Ruseva, V., Dinolov, O.: Energy efficiency of induction motor drives: state of the art, Analysis and Recommendations. Energies 16(20), 7136 (2023). https://doi.org/10.3390/en16207136

Intelligent Data Analysis for Materials Obtained Using Selective Laser Melting Technology

Dmitry Evsyukov[1]([⊠]) [iD], Vladimir Bukhtoyarov[1] [iD], Aleksei Borodulin[1] [iD], and Vadim Lomazov[2,3] [iD]

[1] Bauman Moscow State Technical University, Moscow, Russia
evsjob@yandex.ru
[2] Belgorod State University (BSU), Belgorod, Russia
[3] Belgorod State Agricultural Univerisity Named after V. Gorin Belgorod, Belgorod, Russia

Abstract. In this study, we present a software solution (toolkit) for intelligent data analysis obtained using the selective laser melting (SLM) technology. We have developed a program that uses Data Science approaches and machine learning (ML) algorithms for analyzing and predicting the mechanical properties of materials obtained using the SLM method. The program was trained on a large dataset of SLM materials and was able to achieve an accuracy of 98.9% in terms of the average particle size, using a combination of crystal plasticity and finite element methods (CPFEM) for the Ti-6Al-4V alloy. It allows predicting mechanical properties, such as yield strength, ductility, and toughness, for the structures of Ti-6Al-4V and AlSi10Mg alloys. The study proposes an approach to intelligent data analysis of properties and characteristics of various materials obtained using the SLM technology, based on a formed multidimensional digital model of processes using the developed software solution. The developed set of technologies for intelligent data analysis aimed at optimizing the SLM process demonstrates the potential of machine learning algorithms for improving understanding and optimization of materials obtained through additive manufacturing technologies. Overall, our research emphasizes the importance of developing intelligent solutions for data analysis in materials science and engineering, especially for additive manufacturing technologies such as SLM. By using the developed toolkit that applies machine learning algorithms, specialists can minimize technological production and implementation costs up to 1.2 times, by optimizing the processes of designing and developing materials for various applications, from aerospace industry to biomedical engineering.

Keywords: Selective Laser Melting · Intelligent Data Analysis · Machine Learning · Data Science · Additive Manufacturing Technologies · SLM DS Framework

1 Introduction

The additive manufacturing technology, also known as 3D printing, has revolutionized the way products are produced. Among various additive manufacturing techniques, selective laser melting (SLM) is one of the most promising technologies for producing

V. Jordan et al. (Eds.): HPCST 2023, CCIS 1986, pp. 248–260, 2024.
https://doi.org/10.1007/978-3-031-51057-1_19

complex geometries with high part accuracy from metal powders. However, the quality and properties of the final product strongly depend on the process parameters and the properties of the raw materials used [1].

To address this issue, a suite of data analytics technologies has been developed to optimize the SLM process and predict the properties of the final product. In this article, we present an intelligent solution for analyzing material data obtained during the SLM process.

The developed solution allows for the collection, storage, and processing of data, including external and/or unstructured data, using artificial intelligence algorithms. This enables a better understanding of the processes involved in SLM and increases the accuracy of predicting the properties and quality of the final product.

The solution employs various artificial intelligence methods and algorithms, such as machine learning, statistical analysis, and clustering algorithms, to discover hidden patterns and predict material properties based on the raw material composition and process parameters. Furthermore, the developed solution also allows for the optimization of the SLM process by modeling conditions and predicting optimal process parameters [2] to achieve desired quality and performance criteria. This reduces the time spent on technological stages by automating the process and optimizing quality management, significantly increasing the efficiency of the SLM process.

The aim of this work is to investigate the potential of intelligent data analysis for materials obtained by SLM technology using the developed software solution and to evaluate its effectiveness. The research hypothesis is that the developed software solution will significantly improve the quality and conformity of the sample to the specified characteristics and reduce the stages of the technological process by implementing the prediction of the properties of the obtained material, which, in turn, can lead to improvements in the design and production process of products manufactured by SLM technology.

To achieve the set goal, the following tasks were addressed in this work:

- development of a software solution for collecting, storing, and processing data on materials obtained using SLM technology, using artificial intelligence algorithms;
- analysis of the effectiveness of the developed software solution based on data obtained in experiments with materials obtained using SLM technology.

2 Materials and Methods

To implement intelligent data analysis of the physical, mechanical, chemical, and technological properties of samples grown in different spatial arrangements within a selective laser melting chamber, with the aim of creating an intelligent database of material properties obtained through the selective laser melting process, a software solution was developed for data collection, storage, and processing, including external and/or unstructured data, using artificial intelligence algorithms [3]. For this purpose, a materials property database was created to support and collect information on metallic, polymeric, and ceramic materials at all stages of production, as well as to compare the characteristics obtained during material and/or structure use [4].

Metallic, polymeric, and ceramic materials produced using the SLM process were selected for the study. The description of the studied objects includes characteristics such as size, shape, structure, composition, and other properties.

The quality of the initial components is critical for predicting the mechanical, physical, chemical, and technical characteristics of the obtained samples. In this study, only high-quality components that meet all SLM technology requirements were used.

All data were collected at all stages, starting from the manufacturing stage, with the fixation of input parameters based on the applied source material and manufacturing parameters, and ending with the testing stage, including the methods and techniques of testing and the physical, chemical, mechanical, and other characteristics obtained, with the accumulation of this information in the materials property database.

The materials property database enables data collection from various sources using appropriate connectors, including open, closed, and external sources, and includes the following sections: sample acquisition modes, composition of source components, sample microstructure, stress-strain diagram, ultimate strength, elasticity, Young's modulus, endurance limit, physical characteristics (thermal conductivity, density), technological characteristics (bendability, cutting), impact toughness, hygroscopicity, gas permeability, gas evolution, and others.

The tools and techniques used for data collection and analysis include various data analysis methods, statistical analysis, and analysis of the obtained results.

The main method of data analysis in this study is machine learning and artificial neural networks, which allow for the creation and use of pre-trained machine learning models for predictive analysis of technological parameters to obtain a sample with specified characteristics.

Consider various ML approaches including neural networks, ensemble approaches, Image Processing Methods, clustering algorithms, classification and regression analysis which have proved to be a very useful tool in the study for material analysis.

Multilayer neural networks (MNN) trained on extensive datasets including experimental results and SLM process parameters have proved to be a highly effective tool. They facilitate the identification of complex and hidden relationships between various input variables and material structure. When investigating the accuracy of the model, we included ensemble methods in terms of combining several ML models: Random Forest, Gradient Boosting and numerical method Concurrent Processing Finite Element Model. This allowed us to gain a deeper insight into the processes underlying the creation of materials using SLM technology.

In addition, genetic algorithms and evolution-based optimization have been investigated - these methods can be used to find the best SLM process parameters based on given target material characteristics.

In addition to neural networks, image processing methods have also been actively introduced, which are designed to analyze microscopy of materials obtained during experiments. They help to identify and characterize structural features of the material, such as pores, microstructures and defects. The information obtained has proven to be very valuable in assessing the influence of material structure on its properties.

In addition, various clustering, regression and classification algorithms have been investigated in order to categorize different types of materials and to identify patterns depending on the process parameters, depending on the problems under consideration:

- to study the properties of materials obtained using SLM technology over time and predict their future values, time series analysis may be used;

- to determine the degree of correlation between various characteristics of composite materials and identify the factors that have the greatest influence on their properties to uncover hidden or explicit dependencies, correlation analysis is used;
- to study individual groups of materials based on their characteristics and determine which properties are characteristic of each group, classification algorithms are used.

When performing intelligent analysis of heterogeneous data sets of physical, mechanical, chemical, and technological properties of samples grown [5] at different spatial locations in the growth chamber, the following steps must be taken:

- data storage – this is the first step in intelligent data analysis, which includes developing a method of storing data that will be used in the study;
- data validation – this step involves checking the data for reliability and correctness. It allows you to exclude possible errors and ensure that the data is ready for use;
- data pre-processing and analysis – this step includes data processing to prepare them for use in the analysis, including cleaning the data from outliers, converting the data into a convenient format, and/or selecting a subset of features for further analysis;
- training – at this stage, machine learning models are trained on the processed data.

After training the machine learning models, they can be tested on data sets that were not used in the training process. Various approaches, such as cross-validation, data sampling, and others, can be used to improve the quality of the machine learning models. In cross-validation, the model is tested on multiple data sets, which allows for a more accurate assessment of its quality. Stratified data sampling may be used to consider the distribution of classes in the sample. If the model shows high accuracy and meets the criteria set for testing, this confirms its suitability for use in predicting material properties, classifying materials based on their properties, and identifying dependencies between different properties.

To conduct the study, a sample of 100 specimens was selected for static testing, which were produced using the SLM technology from the following materials: polyamide P2200, steel 316L, titanium alloy Ti6Al4V, and aluminum alloy AlSi10Mg. The test objects were characterized by material properties such as strength, elongation, elasticity, min-max bending, thermal properties, and others. Qualitative input components, as well as data on manufacturing process parameters, were used to predict [6] these characteristics. The specimens were made from different materials and with different manufacturing parameters, including melting temperature, printing speed, size, and shape of the specimen [7].

Various testing instruments and techniques were used to collect data on the materials being developed, such as strength, stability, and wear analyses under different operating conditions. Standard testing methods, such as tensile and bending tests, were used to measure mechanical characteristics. Differential scanning calorimetry (DSC) [8] and thermogravimetry (TGA) [9] methods were used to measure thermal properties. X-ray diffraction (XRD) [10] and scanning electron microscopy (SEM) [11] methods were used to study physical characteristics of the materials. Data was recorded at all stages, from manufacturing with input parameter fixation to testing stages, including testing methods and obtained physical, chemical, mechanical, and other characteristics [12].

To compare the obtained results with the data obtained during the operation of the material and/or constructions, standard testing and result comparison methods [13] were used. One of them is comparison with reference values, which serve as a reference point for comparing the obtained results [14]. These can be formed from standard regulatory technical requirements for materials/constructions, such as GOST, as well as from previous research. Comparison with reference values allows determining whether the obtained results meet the requirements and whether the research goals have been achieved. Another method is comparison with previous data on similar materials. This approach allows comparing obtained results with those obtained in previous studies, which helps to test hypotheses about the results of the research, as well as to identify any differences or unexpected results [15]. Comparison with data obtained from other testing methods is also used to compare results obtained using different methods to confirm the adequacy of testing methods and the possibility of replicating results in other conditions. These methods can be used together or separately to compare results and verify the accuracy of obtained data. They can also be used to assess the quality and accuracy of machine learning models trained on data obtained from experimental tests.

3   Results and Discussion

A solution for investigating materials manufactured using the SLM technology utilizes several ML approaches (Neural Networks, Image Processing Methods, Clustering and Classification Algorithms, and Regression Analysis), including two stages: defining the structure of input experimental and/or modeling data and predicting properties and process parameters that are necessary for understanding the fundamental material properties and fine-tuning the equipment. Exploring the correlation between the structure and properties of materials requires data collection that contains a feature space covering atomic and/or local structural descriptors. These descriptors are assumed to reflect characteristic physical mechanisms and microstructural details for the existing set of materials. They can include atomic properties of elements and atomic fractions in multi-component materials, as well as various indicators such as lattice distortion and local crystallography in metallic systems. This requires constant access to sufficient volumes of data, which include data obtained from high-performance modeling, extensive experiments, mechanical testing and microscopy, as well as from existing material databases and other heterogeneous sources.

The structure of the SLM material investigation process using the developed intelligent analysis solution (see Fig. 1).

Thus, the developed solution allows obtaining new knowledge and information from material properties databases, which can be used to improve the production process, develop new materials, and create innovative technologies.

Studying the relationship between material structure and properties requires collecting data that contains a feature space covering atomic and/or local structural descriptors. These descriptors presumably reflect characteristic physical mechanisms and microstructural details for the existing set of materials. This may include atomic properties of elements and atomic fractions in multicomponent materials, as well as various indicators such as lattice distortion and local crystallography in metallic systems. The "Experiment Settings" window (see Fig. 2).

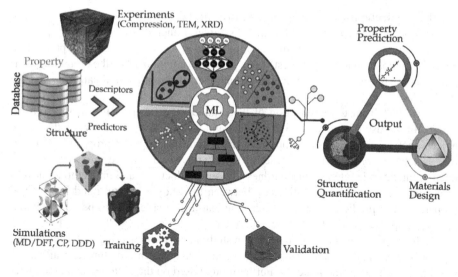

Fig. 1. The SLM material analysis process using machine learning.

To record the results of sample analysis, the "PropertiesOfSamples" table is used, which stores the properties obtained from the analysis of each specific sample. After fixing the property values, they can be changed for a particular sample during subsequent analyses. Additionally, in the "Sample Production" section, there is the "Condition" table that contains information about the state of a specific manufactured sample at a given time. For instance, it can indicate that the sample is in the process of testing or has already passed a check for certain properties.

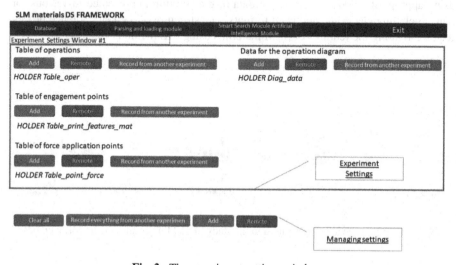

Fig. 2. The experiment settings window.

The "ProductionBatches" table is used to control the production of samples, storing information about the production batches of samples. It includes data such as the production date, the number of manufactured samples, and other necessary data for production control. Moreover, the "SampleRest" table stores information about the remaining samples, such as the date of manufacture, the number of manufactured samples, and the stock balance.

As evident from the description of these tables, the functionality of the solution allows for the use of a combination of local and global attributes to form a dataset of objects that can be analyzed to optimize the volumetric physical properties of materials [1]. For example, with a dataset of metal alloys, we can use information about internal interatomic interactions, hardness, elasticity, ductility, and impact toughness to determine the quantitative relationship between the alloy's structure and its physical properties. The predictive aspect of machine learning enables this to be done without prior assumptions, solely based on existing data.

Such an approach plays a crucial role in studying new engineering materials, allowing for a quick and accurate assessment of how changes in structure can affect the physical properties of the material. Furthermore, based on the obtained data, the material's structure can be optimized to achieve specific physical characteristics. Overall, the well-designed structure of tables and entities allows for efficient management of the sample production and analysis process, quality control, and research to improve material properties and production processes. Therefore, intelligent data analysis enables the acquisition of new knowledge and information from the material properties database that can be used to improve production processes, develop new materials, and create innovative technologies.

The solution uses several ML approaches, including supervised, unsupervised, and reinforcement learning methods, to analyze data. These methods allow for the determination of material properties, the prediction of their properties, and their management using appropriate models. The output data of the solution is presented in various formats, including electronic and print formats. The structure of the Data Science process for investigating the features of SLM materials using ML methods (see Fig. 3).

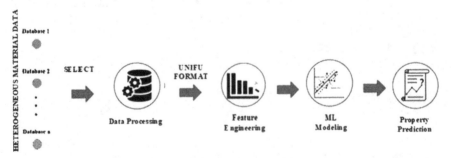

Fig. 3. Data Science approaches used in the study of SLM material traits by ML methods.

In addition, the system allows retraining existing machine learning models on added or existing datasets, providing the ability for more accurate predictions and optimization of material properties based on new data. The user interface provides a wide range of tools for flexible customization of the machine learning pipeline in modeling and research. It includes various tools such as connectors, visualization, preprocessing, models, saving reports to an external file, and evaluators. The number of pipeline steps is unlimited, and multiple visualizations or preprocessings [5] can be used in succession. When selecting a tool in the panel, the available parameters specific to that tool change. The list of tools is stored in a database (DB), where the module name, user name, and brief description of the method are stored. Python script files are stored under the same name.

Tools of the same type have the same input and output format, which allows for easy scaling of the number of tools as needed and according to the scale of the problem being solved. All steps are saved in the pipeline table and will be available for further editing. It is also possible to fix the final model by assigning it a necessary name and using it for quick predictions at any time. The workspace of the intelligent analysis pipeline of LSM materials (see Fig. 4).

The final generated model represents a single Python script that combines all data operations for a specific pipeline. The data should be passed to this model [1] in the format it was received from the connector/source. All data preprocessing will be performed by the script, which will output the results. This approach is a universal one, combining data analysis and DS methods, ensuring easy scalability and flexibility for solutions.

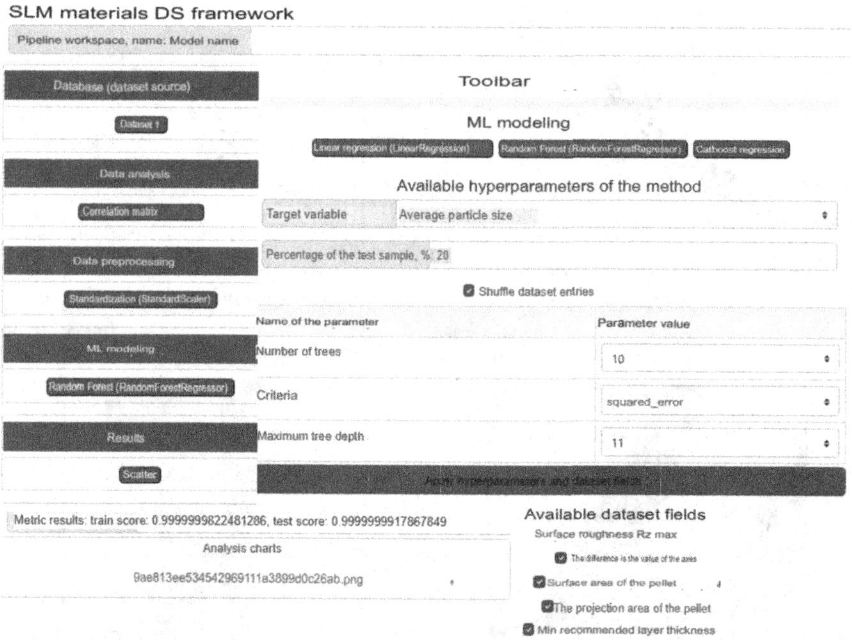

Fig. 4. The working area of the pipeline of intelligent analysis of LSM materials.

To evaluate the quality of prediction models, data on the operation of obtained materials and/or constructions was used. The research results showed that the use of intelligent data analysis can improve the quality of obtained materials and accelerate the process of their creation.

When it comes to SLM metals, volumetric elasticity is one of the functional properties that largely depends on electronic structures, chemical bonds, atomic-molecular arrangement, production history, and thermal treatment. This makes material design a complex task for materials scientists.

In our experiment, data analysis was performed to predict stresses during the tension of metallic materials, such as aluminum alloys, steel, and Ti-6Al-4V alloy at different temperatures, deformation rates, and applied deformations. To achieve accurate assessments of yield strength, hardness, ductility, elasticity, fatigue properties, and fracture toughness, a neural network [1] was used, which was trained to find the optimal alloy composition in combination with thermal treatment parameters.

In addition, intensive research work has been carried out in the field of amorphous metals. The modeling graph and yield strength evaluation of the composition were conducted with the use of solid solution strengthening structure in accordance with the specified strength properties that depend on the composition. Empirical measurements combined with a relevant set of parameters [16] were used to tune the parameters of the gradient boosting trees algorithm (see Fig. 5).

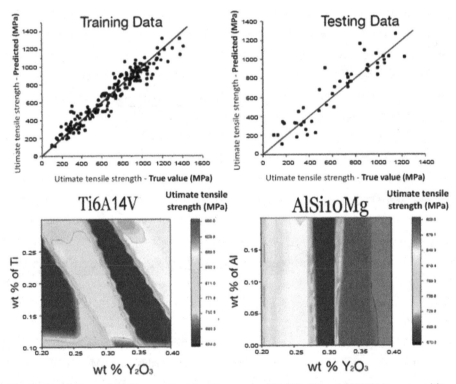

Fig. 5. Simulation and yield strength estimation graph of Ti-6Al-4V and AlSi10Mg composition.

In this research, we conducted experiments aimed at investigating crystal deformation in polycrystalline metals and alloys, with a focus on Ti-6Al-4V material. Our solution also allows the investigation of Digital Image Correlation (DIC) and Multiscale Material Modeling (MI) techniques combined with machine learning.

Initially, samples of Ti-6Al-4V material were prepared for subsequent analysis. These samples were subjected to experiments using a variety of previously reported techniques, including ultra-high resolution electron backscatter diffraction (EBSD) and surface profilometry. EBSD provides information about the microstructure of the crystals in the sample, while profilometry allows the measurement of surface strain.

The next step was to create models to analyze the deformation in the Ti-6Al-4V material. For this purpose, crystal plasticity and finite element modeling (CPFEM) techniques were used to create mathematical models to predict the deformation behavior of the material.

The next aspect of the study was the correlation between the data obtained from experiments (EBSD and profilometry) and the results predicted by CPFEM methods. In this step, a comparison between the deformation fields obtained from CPFEM models and real deformation measurements on the specimens was performed [17].

In addition, neural networks were employed to analyze and interpret the data obtained by DIC and MI in order to trace complex patterns and dependencies in the data, which in the final solution improved the accuracy of the analysis and provided a better understanding of the crystal deformation processes in Ti-6Al-4V material. The neural network was organized as a Convolutional Neural Network (CNN), which is a typical choice for image analysis.

The input layer accepts images obtained from DIC and MI methods. The convolutional layers were used to extract features from images (detect patterns and structures in deformed materials). Full-link layers accept the extracted features and performed analysis to further interpret the data. The output layer generates the analysis results including information about deformation and material behavior.

To optimize the neural network structure and its hyperparameters, we used a genetic algorithm. This method allowed us to systematically modify the network architecture, select optimal activation functions, the number of layers and their sizes, and tune other parameters to achieve the best results in analyzing deformation data.

Thus, very good agreements were obtained, and the solution demonstrates the application of CPFEM methods for studying plasticity and deformation in Ti-6Al-4V alloy with the use of ML (see Fig. 6).

Overall, the use of various data collection and analysis methods with the developed software solution toolkit has allowed for a more complete and accurate understanding of the properties of materials produced by SLM technology and to determine optimal manufacturing [18] parameters to achieve the required characteristics with sufficiently high levels of accuracy and efficiency in designing new SLM materials and products made from them.

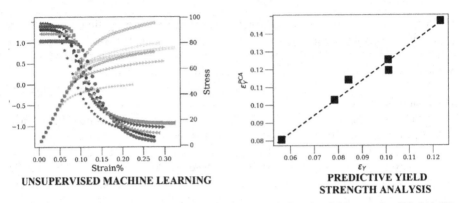

Fig. 6. A software environment for investigating and predicting the yield strength of Ti-6Al-4V, using ML methods.

4   Conclusions

This article is an intermediate step towards achieving the ultimate goal. As a result of the study, a software solution for intelligent data analysis of materials produced using SLM technology has been developed. This solution has significantly increased the efficiency and accuracy of the analysis, as well as revealed new patterns in the data. Within the study, experiments were conducted on a dataset obtained from studies of the structure of Ti-6Al-4V and AlSi10Mg alloys, (see Fig. 4, 5 and 6), and machine learning and data analysis methods were applied to determine the optimal parameters for the yield strength of the SLM process.

The results showed that our software solution enabled achieving a prediction accuracy of 98.9% for mechanical properties of materials obtained through SLM technology, in terms of the average particle size, by combining crystal plasticity and finite element methods (CPFEM) for the Ti-6Al-4V alloy, which surpasses the accuracy of previous studies in this field. Additionally, our solution allows reducing data analysis time by several times, significantly simplifying the analysis process and increasing the efficiency of using SLM materials in various application areas [19], through more efficient SLM process design using machine learning methods. Thus, our software solution can be further used for analyzing data for other materials obtained through SLM technology and for improving the design and production process of materials with desired properties.

Despite significant achievements, there are some limitations to our study. Firstly, we used data only from one source - the "https://viam.ru/review/5942" database. This may limit the generalization of results to other SLM materials that may have different compositions and properties. Additionally, our model does not consider the influence of other factors on material properties, such as ambient temperature, humidity, etc. Further research is necessary for a more complete understanding of the interaction between SLM process parameters and material properties. It should also be noted that the obtained results may be dependent on specific SLM process conditions and the data used, which will allow us to expand our model in the future to include additional factors, such as temperature and humidity, as well as use data from other sources [1].

References

1. Qi, X., Chen, G., Li, Y., Cheng, X., Li, C.: Applying neural-network-based machine learning to additive manufacturing: current applications, challenges, and future perspectives. J. Eng. **5**(4), 721–729 (2019)
2. Wang, Y., Chen, X., Jayalakshmi, S., Singh, R.A., Sergey, K., Gupta, M.: Process parameters, product quality monitoring, and control of powder bed fusion. In: Chen, S., Zhang, Y., Feng, Z. (eds.) Transactions on Intelligent Welding Manufacturing. TIWM, pp. 89–108. Springer, Singapore (2020). https://doi.org/10.1007/978-981-13-8192-8_4
3. Chen, Y., Wang, H., Wu, Y., Wang, H.: Predicting the printability in selective laser melting with a supervised machine learning method. Materials **13**(22), 5063 (2020)
4. Barantsov, I.A., Pnev, A.B., Koshelev, K.I., Tynchenko, V.S., Nelyub, V.A., Borodulin, A.S.: Classification of acoustic influences registered with phase-sensitive OTDR using pattern recognition methods. Sensors **23**(2), 582–594 (2023)
5. Bukhtoyarov, V.V., Tynchenko, V.S., Nelyub, V.A., Masich, I.S., Borodulin, A.S., Gantimurov, A.P.: A study on a probabilistic method for designing artificial neural networks for the formation of intelligent technology assemblies with high variability. Electronics **12**(1), 215–229 (2023)
6. Mikhalev, A.S., et al.: The orb-weaving spider algorithm for training of recurrent neural networks. Symmetry **14**(10), 2036–2048 (2022)
7. Masich, I.S., Tyncheko, V.S., Nelyub, V.A., Bukhtoyarov, V.V., Kurashkin, S.O., Borodulin, A.S.: Paired patterns in logical analysis of data for decision support in recognition. Computation **10**(10), 185–197 (2022)
8. Zhu, Y., Peng, T., Jia, G., Zhang, H., Xu, S., Yang, H.: Electrical energy consumption and mechanical properties of selective-laser-melting-produced 316L stainless steel samples using various processing parameters. J. Clean. Prod. **208**, 77–85 (2019)
9. Wang, X., Zhang, C.H., Cui, X., Zhang, S., Chen, J., Zhang, J.B.: Novel gradient alloy steel with quasi-continuous ratios fabricated by SLM: material microstructure and wear mechanism. Mater. Character. **174**, 111020 (2021)
10. Li, J., Cao, L., Hu, J., Sheng, M., Zhou, Q., Jin, P.: A prediction approach of SLM based on the ensemble of metamodels considering material efficiency, energy consumption, and tensile strength. J. Intell. Manuf. **33**, 1–16 (2022). https://doi.org/10.1016/j.matchar.2021.111020
11. Wegener, K., Spierings, A., Staub, A.: Bioinspired intelligent SLM cell. Procedia CIRP **12**(1), 88–102 (2020)
12. Zhao, T., Zhang, S., Zhou, F.Q., Zhang, H.F., Zhang, C.H., Chen, J.: Microstructure evolution and properties of in-situ TiC reinforced titanium matrix composites coating by plasma transferred arc welding (PTAW). Surf. Coat. Technol. **424**(504), 127637–127649 (2021)
13. Yin, T.Y., Zhang, S., Zhou, F.Q., Huo, R.J., Zhang, C.H., Chen, J.: Effects of heat treatment on microstructure and wear behavior of modified aluminum bronze coatings fabricated by laser cladding. J. Mater. Eng. Perf. **31**(6), 4294–4304 (2022)
14. Yadav, P., Rigo, O., Arvieu, C., Le Guen, E., Lacoste, E.: In situ monitoring systems of the SLM process: On the need to develop machine learning models for data processing. Crystals **10**(6), 524 (2020)
15. Barrionuevo, G.O., Ramos-Grez, J.A., Walczak, M., Betancourt, C.A.: Comparative evaluation of supervised machine learning algorithms in the prediction of the relative density of 316L stainless steel fabricated by selective laser melting. J. Adv. Manuf. Technol. **113**(1), 419–433 (2021)
16. Wang, Z., Xiao, Z., Tse, Y., Huang, C., Zhang, W.: Optimization of processing parameters and establishment of a relationship between microstructure and mechanical properties of SLM titanium alloy. Opt. Laser Technol. **112**(5), 159–167 (2019)

17. Franczyk, E., Machno, M., Zębala, W.: WInvestigation and optimization of the SLM and WEDM processes' parameters for the AlSi10Mg-sintered part. Materials **14**(2), 410–424 (2021)

18. Papazoglou, E.L., Karkalos, N.E., Karmiris-Obratański, P., Markopoulos, A.P.: On the modeling and simulation of SLM and SLS for metal and polymer powders. Arch. Comput. Methods Eng. **1**(3), 1–33 (2021)

19. Cao, L., Yuan, X.: Study on the numerical simulation of the SLM molten pool dynamic behavior of a nickel-based superalloy on the workpiece scale. Materials **12**(14), 2272–2284 (2019)

Computing Technologies in Information Security Applications

Fourier Chromagrams for Fingerprinting, Verification and Authentication of Digital Audio Recordings

Andrey Lependin$^{(\boxtimes)}$ (ID), Pavel Ladygin (ID), Valentin Karev (ID),
and Alexander Mansurov (ID)

AltSU – Altai State University, Lenin Ave. 61, 656049 Barnaul, Russia
andrey.lependin@gmail.com

Abstract. In this paper, a new approach for calculating binary audio fingerprints was proposed. This approach was based on the analysis of Fourier chromagrams obtained from the processed music recordings (audio files). The calculated binary audio fingerprints allow for bit-by-bit matching and comparison of original and modified music recordings. For performance testing, a dataset of over 50 original recordings of music played on a variety of instruments using different playing techniques was collected. In addition, distorted versions of the original recordings with altered tempo and realistic additive noise were produced and added to the test dataset. Calculations of similarity values between different audio fingerprints within the same groups of music recordings help reveal the expected robustness of the proposed approach against the possible distortions mentioned earlier. The impacts of distortions on chromagrams and calculated audio fingerprints were thoroughly analyzed and discussed in the paper. The median values of the similarity between the original and distorted recordings were found to be greater than 85%. The proposed approach proves to be quite useful in real life forensic studies and tasks of verification and authentication of music pieces and recordings.

Keywords: Digital Fingerprint · Musical Audio Signal · Spectrogram · Fourier Chromagram · Forensic Musicology · Audio Recordings Authentication

1 Introduction

Digital fingerprinting is an important part of modern audio recordings identification technologies. These technologies can be implemented as a stand-alone program solution or as plug-in modules of information systems that automatically analyze the music broadcasts, arrange music collections in the social media, and perform instant identification of music content.

The most basic requirements for digital fingerprinting methods are the following:

- computational simplicity;
- small fingerprint size for easy storing, matching, and retrieval;
- invariance to small distortions and changes;

© The Author(s), under exclusive license to Springer Nature Switzerland AG 2024
V. Jordan et al. (Eds.): HPCST 2023, CCIS 1986, pp. 263–275, 2024.
https://doi.org/10.1007/978-3-031-51057-1_20

- simple and guaranteed discrimination or comparison with other fingerprints.

The most common approaches to digital audio fingerprinting are the well-known method designed by the Philips company [1], Shazam's identification technology [2, 3], and the number of methods developed and used by the Google company [4]. Typically, digital audio fingerprinting methods are based on the Fourier spectral analysis followed by processing of spectral characteristics [3, 5], calculation of mel-frequency cepstrum coefficients [6] or use of wavelet transform [7].

Digital audio fingerprints can also be utilized for the highly sophisticated, albeit crucial, forensic studies, or forensic musicology. In some cases, proving illegal use of audio recordings or copyright infringement [8, 9] requires verification of the original music composition or plagiarism detection in a certain amount of music compositions.

The conventional approaches to the forensic musicology during litigations include the analysis of a music composition aurally (by ear) along with the comparison of music scores. The key factors for the experts here are the tonal matching (or mismatching), the comparison of pitch and intensity of musical notes, and the analysis of the consonance of musical notes [9, 10]. With few exceptions, the use of technical means is limited to comparing music recordings by overlaying one recording on another and matching the metadata of the compared music audio files. Thus, a huge influence of subjectivity, expert's qualification and experience play a crucial part in the forensic study. Unfortunately, some techniques used at the recording and production stages of music compositions (such as key and rhythm changing, adding artificial distortions and sound effects) often cannot be identified and overcome by experts, leading to inaccurate conclusions.

Transforming the spectral representation of the audio signal into a chromagram representing the note sequence appears to be a convenient way to analyze musical recordings. A detailed analysis was conducted to understand the effectiveness of chromagrams based on the Fourier transform [11] and constant-Q transform [12]. The chromagrams were obtained for the same piece of music played by different instruments and for the piece of music after transposition. Fourier-based chromagrams have been shown to exhibit sufficient invariance and robustness to possible changes and transformations introduced in the musical composition.

This paper presents a novel approach for computing digital audio fingerprints of instrumental audio recordings, suitable for forensic investigations and comparison of musical compositions. The proposed approach uses Fourier-based chromagrams [11, 13]. In addition to computational simplicity, this approach allows analyzing and comparing music compositions and recordings, similar to the common methods used by experts during their studies to authenticate musical pieces.

2   Proposed Approach

The proposed procedure for verification and authentication of music audio recordings uses direct comparison of digital audio fingerprints (Fig. 1). Two audio recordings, S_1 and S_2, are channeled into the input of the procedure. The recording S_1 stands for the original music composition, and the recording S_2 stands for the alleged alternative version. They are split into a sequence of frames of equal length (90 ms). Each frame is passed through the Hann window filter and then processed by the short-time Fourier transform (STFT)

[11]. The obtained Fourier spectra of the processed frames are used to calculate the Fourier-based chroma features [13] for each audio recording. The chromagrams display the relative notes' intensity when all note pitches are reduced down to a single octave (12 possible values: C, C♯, D, D♯, E, F, F♯, G, G♯, A, A♯, B). Next, all peaks that correspond to the specific note pitches are detected and binarized within each frame. The transition between note pitches is encoded as follows: "01" – when pitch height increases, "10" – when pitch height decreases, "00" – when pitch height remains unchanged.

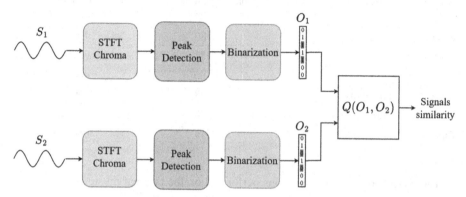

Fig. 1. The proposed procedure to compare music audio recordings.

The resulting binary sequences O_1 and O_2 have their length proportional to the length of the audio recordings at the input. They are considered as the audio fingerprints of the corresponding audio recordings. It is obvious that mismatching melodic patterns lead to mismatching binary sequences. Here, in the example in Fig. 1, the mismatching bits are marked with red. The value $Q(O_1, O_2)$ is calculated to estimate the similarity of the binary sequences O_1 and O_2:

$$Q(O_1, O_2) = (1 - N(O_1, O_2)/N_0) \cdot 100\%, \tag{1}$$

Where $N(O_1, O_2)$ is the number of matching bits in two sequences, and N_0 is the length of the shortest sequence (the number of all bits in the sequence).

3 Dataset for Testing

A special dataset is arranged to estimate the performance and effectiveness of the proposed approach. The dataset consists of three types of music compositions (Table 1) that include audio recordings of original music compositions played on different instruments, as well as audio recordings of altered music compositions. The alterations are done by adding background noise of different power or changing the tempo of the original music.

There are 52 audio files (recordings) of original music compositions (original samples). The selected audio files have a length from 3 s to 216 s and a sampling frequency of 44.1 kHz. Music compositions are played on different instruments, which results in different corresponding spectral ranges. Wind instruments produce sounds with a high-frequency spectrum (from 1 kHz and higher). A wider spectral range (from 40 Hz and higher) is common for string and keyboard instruments. Also, music compositions with and without chords (no chords) are included.

Table 1. The dataset structure.

Instrument	Music Chords	Original Samples	Samples with Altered Tempo	Samples with Additive Noise
Wind	no chords	10	80	400
Keyboard	no chords	19	152	760
	with chords	12	96	480
String	no chords	6	48	240
	with chords	5	40	200
Total	–	52	416	2080

There are two types of alterations performed on the original music compositions. The first one is to introduce slight changes to the tempo. Typically, it is the most common alteration performed by the "so-called" pirates. Here, the original tempo is sped up or slowed down by 1%, 2%, 3%, and 10%, producing eight different versions for each original music composition. The second type of alteration plays a crucial role for testing the stability and effectiveness of the proposed approach. It assumes adding background noise to the original music compositions. Noise fragments taken from the Audioset [14], Freesound [15], and DEMAND [16] datasets are combined with the original music compositions. Various types of noise fragments imitating the voices of people, the sounds of music instruments, and urban and nature sounds are utilized to simulate real-life situations during the music recording process. Noise fragments spectra are non-stationary and include the whole range of audio frequencies. The amplitudes of each noise fragment are adjusted to have the resulting signal-to-noise (SNR) ratio of the combined recording within the range of -5 dB to 30 dB, with a step of 5 dB.

4   Numerical Experiments and Discussion

4.1   Influence of Changes in Tempo

To estimate the influence of changes in tempo on the calculated audio fingerprint, the value Q is calculated for fingerprints of each original sample and its alterations with different tempos. The histogram of calculated similarity values Q for the testing dataset is shown in Fig. 2. Table 2 contains the numerical median values of Q for convenience.

The obtained values of Q demonstrate that changes in tempo rarely lead to severe distortions and drastic changes in the similarity of calculated fingerprints. The distortions appearing in chromagrams are explained by several factors. Typically, speeding up or slowing down the tempo results in changes of the frequencies of the primary

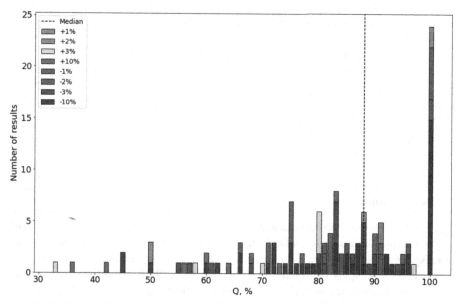

Fig. 2. The similarity between original samples and their versions with altered tempos.

and secondary harmonics of sounds produced by instruments. It can cause the incorrect detection of notes in the spectrum and the incorrect encoding of the transitions between note pitches. However, changes of note frequencies are irrelevant because the calculated audio fingerprint is based on the encoded transitions. Figure 2 shows that low values of similarity Q can also be obtained if samples with different changes in tempo (speeding up - orange (+3%) and green (+10%) bars, slowing down – orange (−1%) and blue (−10%) bars) are compared with the originals. According to the values in Table 2, slowing the tempo down results in less distortion than speeding the tempo up. The explanation can rely on the fact that speeding up shortens the pauses (time gaps) between note pitches. Consequently, uncertainty level decreases when calculating the basic frequency of a note at the moment of its decay, and transitions become blurred and less distinctive. Thus, it is harder to detect and encode the transitions between note pitches correctly.

Table 2. Median values of similarity Q between original samples and their versions with altered tempos

Change in Tempo, %	Median Value of Q, %
−10	86.5
−3	86.5
−2	85

(*continued*)

Table 2. (*continued*)

Change in Tempo, %	Median Value of Q, %
−1	95
1	83
2	91
3	85
10	83

4.2 Noise Robustness

Figure 3 demonstrates the dependence of median similarity values on SNR values. The audio fingerprint of each original noiseless sample is compared with fingerprints of all its noisy versions (derivative samples with added noise at the fixed SNR). It is clearly seen that noisy samples with SNR values greater than 20 dB exhibit practically ideal similarity to the noiseless samples. Expectedly, the decrease of the SNR values results in a rapid decrease of the median similarity values Q.

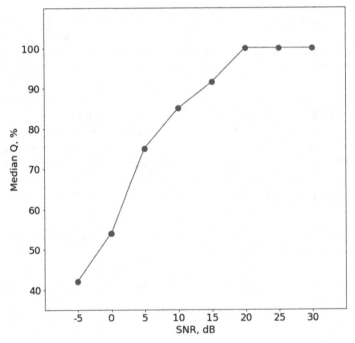

Fig. 3. Dependence of median values of similarity Q between the noiseless (original) and noisy versions of audio recordings on the SNR values of noisy versions.

The key factor that explains the decrease of similarity values Q lies in the uncertainty during the detection of peaks (note pitches) on chromagrams and encoding the transitions between them. An example of how added noise distorts the original noiseless sample is shown in Fig. 4 and Fig. 5. The Fourier spectrogram of the original noiseless sample demonstrates clearly the primary and secondary harmonics of sounds produced by the instrument (Fig. 4a). The corresponding chromagram (Fig. 4b) is also clear and distinctive.

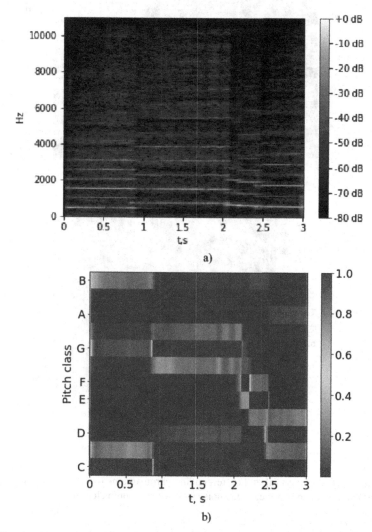

Fig. 4. Fourier spectrogram (a) and corresponding chromagrams (b) of a noiseless sample. Musical instrument is flute.

The added noise distorts the spectrogram significantly (Fig. 5a), and the resulting chromagram (Fig. 5b) contains numerous false note pitches (within the time range from

2 s to 2.5 s) at the moment when the note decays. This leads to detection and encoding of false transitions when calculating the fingerprint. It is evident that preliminary noise removal should be performed on the processed signal to obtain a better fingerprint later.

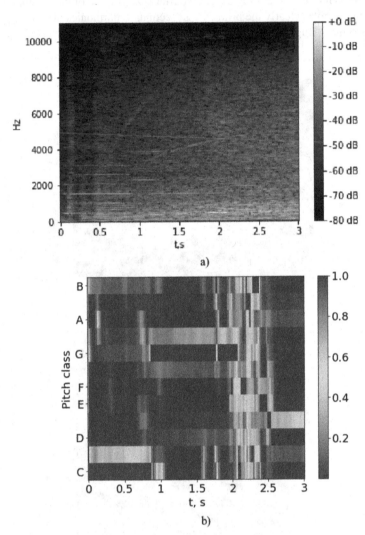

Fig. 5. Fourier spectrogram (a) and corresponding chromagrams (b) of a sample with additive noise. Musical instrument is flute, added noise is of type "vacuum cleaner".

4.3 Influence of Playing Instruments and Music Chords

Table 3 contains median similarity values calculated for samples played by a specific type of music instrument. Similarity values are obtained separately for cases with slowing down/speeding up the tempo and for cases with additive noise. The peculiarities of distorted audio recordings (samples) are not investigated in detail. By following the Table 3 data, the general picture of how features of musical instruments and their ways of producing sounds influence the calculated audio fingerprints becomes evident. Different musical instruments have different timbral characteristics, produce sounds within different frequency ranges, etc. It certainly affects the calculations of audio fingerprints and values of similarity between two audio recordings.

Table 3. Median values of similarity Q between original samples and their versions with two types of alteration.

Instrument	Music Chords	Median Value of Q, %	
		Change in Tempo	Additive Noise
Wind	no chords	87.5	77.0
Keyboard	no chords	93.5	90.0
	with chords	88.0	90.0
String	no chords	83.0	94.0
	with chords	87.5	92.0
All types		88.0	85.0

Audio fingerprints of recordings with music played with wind instruments demonstrate the lowest values of similarity. The arranged testing dataset contains recordings with wind instruments like flutes, saxophones, and pipes. Music melodies are played with the "vibrato" and "glissando" effects. An example of a chromagram with such effects is shown in Fig. 6. There are several ambiguities encountered at the transition areas and areas with slow decay of note pitches.

Playing the music piece with and without chords results in severe changes of the spectrograms and chromagrams of audio recordings. Figure 7 provides an example where the same music piece is played without chords (Fig. 7a) and with chords (Fig. 7b). A lot of additional secondary harmonics and subharmonics appear in the spectrogram, which completely distort the corresponding chromagram (Fig. 7b). The area within the time range from the beginning to 0.5 s reveals instability and no clearly distinctive note pitches. This further complicates the detection of transitions and the further calculation of audio fingerprints.

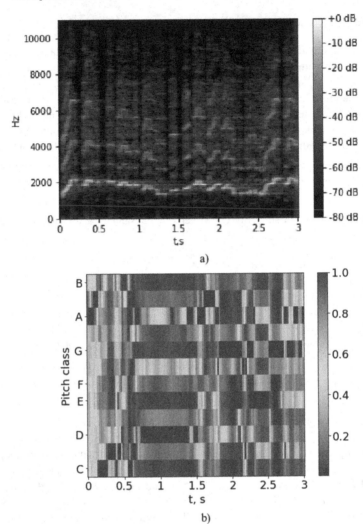

Fig. 6. Fourier spectrogram (a) and corresponding chromagram (b) of a music melody played on a pipe. The following effects were used: "glissando", "vibrato".

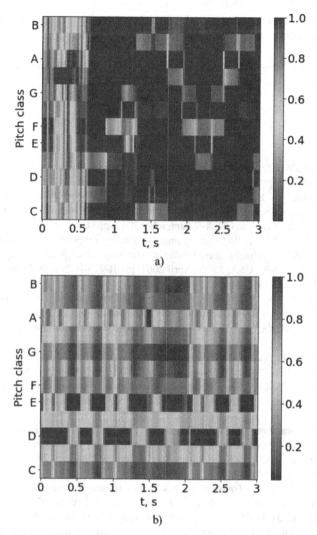

Fig. 7. Chromagrams of the same melody played without chords (a) and with chords (b), respectively. In the first case (a) it was played on the flute, in the second (b) it was played on the guitar.

5 Conclusion

The proposed approach allows the direct comparison of music recordings using their calculated binary audio fingerprints. The approach has been tested on the arranged dataset. It is found that the proposed approach demonstrates its robustness against various potential distortions that can be introduced accidentally (noise addition) or deliberately (changes in tempo). It employs a simple and convenient way to calculate digital audio fingerprints and match two audio fingerprints at any specific place and time within the compared audio recordings. This advantage highlights the potential prospects of the proposed approach being used in studies of forensic musicology during litigations.

Acknowledgements. This work was supported by the grant from the Russian Science Foundation, project no. 22-21-00199, https://rscf.ru/en/project/22-21-00199/.

References

1. Haitsma, J., Kalker, T.: A highly robust audio fingerprinting system. J. New Music Res. **32**(2), 211–221 (2003). https://doi.org/10.1076/jnmr.32.2.211.16746
2. van Nieuwenhuizen, H.A., Venter, W.C., Grobler, L.M.: The study and implementation of SHAZAM's audio fingerprinting algorithm for advertisement identification. In: Proceedings of SATNAC 2011, London, UK, 4–7 September 2011, pp. 1–4 (2011)
3. Baluja, S., Covell, M.: Audio fingerprinting: combining computer vision & data stream processing. In: 2007 IEEE International Conference on Acoustics, Speech and Signal Processing ICASSP'07, Honolulu, USA, 15–20 April 2007, pp. II-213-II-21 (2007). https://doi.org/10.1109/ICASSP.2007.366210
4. Baluja, S., Covell, M.: Content fingerprinting using wavelets. In: The 3rd European Conference on Visual Media Production CVMP 2006, part of the 2nd Multimedia Conference, 2006. 29–30 November 2006, London, UK, pp. 198–207 (2006). doi:https://doi.org/10.1049/cp:20061964
5. Cano, P., Batlle, E., Kalker, T., Haitsma, J.: A review of audio fingerprinting. J. VLSI Signal Process. **41**(3), 271–284 (2005). https://doi.org/10.1007/s11265-005-4151-3
6. Sonnleitner, R., Widmer, G.: Robust quad-based audio fingerprinting. IEEE/ACM Trans. Audio Speech Lang. Process. **24**(3), 409–421 (2015). https://doi.org/10.1109/TASLP.2015.2509248
7. Cancela, P., Rocamora, M., López, E.: An efficient multi-resolution spectral transform for music analysis. In: Proceedings of 10th International Society for Music Information Retrieval Conference (ISMIR 2009), Kobe, Japan, 26–30 October 2009, pp. 309–314 (2009). https://doi.org/10.5281/zenodo.1416788
8. Mopas, M., Curran, A.: Translating the sound of music: forensic musicology and visual evidence in music copyright infringement cases. Can. J. Law Soc./La Revue Canadienne Droit Et Société. **31**(1), 25–46 (2016). https://doi.org/10.1017/cls.2016.4
9. Begault, D.R., Heise, H.D., Peltier, C.A.: Analysis criteria for forensic musicology. In: Proceedings of Meetings on Acoustics, ICA 2013, Montreal, Canada, 2–7 June 2013, vol. 19, p. 060005. Acoustical Society of America (2013). https://doi.org/10.1121/1.4799479
10. Brauneis, R.: Musical work copyright for the era of digital sound technology: looking beyond composition and performance. Tul. J. Tech. Intell. Prop. **17**(1), 1–60 (2014). https://doi.org/10.2139/ssrn.2400170
11. Ladygin, P.S., Lependin, A.A., Mansurov, A.V.: Authentication of music audio recordings using digital fingerprints based on STFT and CQT chromatograms. High-Perf. Comput. Syst. Technol. **7**(1), 46–52 (2023)
12. Brown, J.C.: Calculation of a constant Q spectral transform. J. Acoust. Soc. Am. **89**(1), 425–434 (1991). https://doi.org/10.1121/1.400476
13. Jiang, N., Grosche, P., Konz, V., Müller, M.: Analyzing chroma feature types for automated chord recognition. In: Proceedings of AES 42nd International Conference, Ilmenau, Germany, 22–24 July 2011, pp. 1–10. AES, New York (2011)

14. Gemmeke, J.F., et al.: Audio set: an ontology and human-labeled dataset for audio events. In: 2017 IEEE International Conference on Acoustics, Speech and Signal Processing (ICASSP), New Orleans, LA, USA, 5–9 March 2017, pp. 776–780 (2017). https://doi.org/10.1109/ICASSP.2017.7952261

15. Font, F., Roma, G., Serra, X.: Freesound technical demo. In: Proceedings of the 21st ACM International Conference on Multimedia, 21–25 October 2013, pp. 411–412 (2013). https://doi.org/10.1145/2502081.2502245

16. Joachim, T.J., Ito, N., Vincent, E.: The diverse environments multi-channel acoustic noise database (DEMAND): a database of multichannel environmental noise recordings. In: Proceedings of Meetings on Acoustics, ICA 2013, Montreal, Canada, 2–7 June 2013, vol. 19, p. 035081. Acoustical Society of America (2013). https://doi.org/10.1121/1.4799597

Methodology of Expert-Agent Cognitive Modeling for Preventing Impact on Critical Information Infrastructure

Pavel Panilov[✉] , Tatyana Tsibizova , and Georgy Voskresensky

Bauman Moscow State Technical University, Moscow, Russia
`panilovp.a@bmstu.ru`

Abstract. This scientific article presents a comparative analysis of different threat prediction models used in security systems. Our expert-agent cognitive model has shown to be the most effective across a range of threat levels, with the highest performance compared to other models. The Circular Protection and Life Tree models also showed promising results but are less effective at higher threat levels. However, the Interval Confidence Interval model showed the worst performance in this comparative analysis. We have also identified the optimal input parameter values for our expert-agent cognitive model, which result in a 95% improvement in its prediction accuracy. Our model achieves the highest quality of prediction at Knowledge Base = 100, Expert Rating = 5, and Threats = 500. The performance of our model starts at around 60% accuracy with 50 threats and reaches a peak of 80% accuracy at 200 threats, gradually decreasing at higher threat levels. Our findings indicate that our proposed model is recommended for predicting and warning about potential threats in security systems. Further research can help optimize the parameters of our model for even more effective threat prediction and warning.

Keywords: Expert-Agent Cognitive Modeling · Critical Information Infrastructure · Cybersecurity · Attack Scenarios · Prevention Recommendations · Multi-Criteria Optimization

1 Introduction

In the modern world, critical information infrastructure (CII) plays a crucial role in ensuring the functioning of various spheres of life, including energy systems, telecommunications networks, transportation management, banking and financial systems, healthcare systems, and others. However, the increasing dependence on these systems makes them attractive targets for malicious actors seeking to disrupt or compromise their operations. As a result, there is an urgent need to develop methodologies and models capable of effectively safeguarding CII from potential threats and attacks [1–3].

There are several popular models for protecting CII, such as the circular protection model, the "Tree of Life" model, and the "Interval Confidence Intervals" model. Each of these models has its own limitations and shortcomings.

V. Jordan et al. (Eds.): HPCST 2023, CCIS 1986, pp. 276–287, 2024.
https://doi.org/10.1007/978-3-031-51057-1_21

One of the most promising methodologies for CII protection is expert-agent cognitive modeling (EACM), which integrates expert knowledge and agent-based modeling to analyze and prevent impacts on CII. This article focuses on the application of the EACM methodology for mitigating threats and protecting critical information infrastructure.

Several studies have been conducted to address the protection of CII from various perspectives. For example, Kondratenko and Sidorova [4] discussed the modeling of cybersecurity threats to CII, while Nikolaev and Mukhanov [5] explored the use of machine learning methods for detecting attacks in computer networks. Belyaeva and Stepanova [6] investigated the application of neural networks for prediction and classification tasks, while Zotov, Lysenko, and Nekrasov [7] focused on data analysis methods for intrusion detection.

Considering the importance of CII protection, this article aims to contribute to the existing scientific community by presenting the application of the EACM methodology for preventing impacts on critical information infrastructure. By combining machine learning and cognitive modeling, this methodology offers a powerful toolkit for enhancing the cybersecurity resilience of critical information infrastructure.

2   Analysis of Existing Models

Currently, there are several models designed for preventing impacts on critical information infrastructure (CII). The most well-known models are:

- The Circular Protection Model;
- The Tree of Life Model;
- The Interval Confidence Interval Model.

The Circular Protection Model involves creating a protection system consisting of several protection circles, each of which performs a specific function in protecting the CII [11]. This model has the following disadvantages:

- Implementation complexity;
- Inability to account for all possible threats;
- Inefficiency in the event of new threats.

The Tree of Life Model is a tree where each node describes an event that can occur in the system. This model has the following disadvantages [8, 9]:

- Implementation complexity;
- Inability to account for all possible events;
- Inefficiency in the event of new events.

The Interval Confidence Interval Model is based on using statistical methods to determine the probability of a particular event occurring [10–12]. However, this model also has the following disadvantages:

- The need for constant data updating and analysis;
- Inefficiency in the event of new events;
- The need for a large amount of data to achieve accurate forecasting.

3 Expert-Agent Cognitive Modeling

To address the problem of preventing impact on critical information infrastructure, it is proposed to use the methodology of expert-agent cognitive modeling (EACM). EACM is based on the use of expert knowledge and agent technologies to develop models capable of effectively preventing impact on CI [13, 14]. EACM allows creating models that can adapt to the changing environment and take into account new threats and events [15]. The main components of EACM are:

- An expert system that is used for storing and processing expert knowledge;
- An agent system that is used for modeling system behavior and providing its protection;
- Machine learning methods that are used for data analysis and detecting new threats and events.

3.1 Model for Preventing Impact on Critical Information Infrastructure

We propose a model for preventing impacts on the critical information infrastructure (CII) of a mechanical engineering enterprise using EACM.

- The expert system will contain expert knowledge about potential threats and necessary protective measures [16];
- The agent system will ensure the protection of the enterprise's CII. Agents will be responsible for access control, system monitoring, threat detection, and protective action [17, 18];
- Machine learning methods will be used to analyze data and detect new threats. For example, anomaly analysis can be used to detect unusual behavior in the system.

As a result, the model for preventing impacts on the CII of a mechanical engineering enterprise will have the following advantages:

- Adaptability to changing environments;
- Consideration of all possible threats and events, including new ones;
- Effectiveness in detecting threats and taking protective measures;
- Rapid response to new events and threats;
- Minimization of risks to the CII of the mechanical engineering enterprise.

We propose an algorithm for designing the EACM model for a mechanical engineering enterprise (see Fig. 1).

Designing a model may include the following steps:

- Defining the goals of protecting critical information infrastructure (CII). The first step is to determine the goals of protecting the critical information infrastructure (CII) of the engineering enterprise. It is necessary to analyze the types of CII and identify the threats and impacts that can affect them, as well as the vulnerabilities that can be exploited for attacks. It is also necessary to determine which data and systems of CII are the most critical and require special protection [19];
- Creating expert systems. Next, the creation of expert systems follows, which will be used for monitoring and analyzing the CII. Expert systems can be created based on

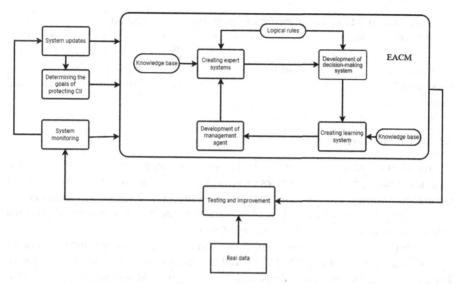

Fig. 1. Model EACM.

knowledge bases, logical rules, and machine learning to automatically analyze data, detect threats, and issue warnings;

- Development of decision-making system. The decision-making system can be created based on logical rules, expert systems, and machine learning. It should determine what measures to take in response to detected threats and impacts on the CII;
- Development of management agent. The development of a management agent involves creating a system that can manage the information systems of a mechanical engineering enterprise, including rebooting systems, changing settings and configurations, as well as taking other necessary measures to eliminate threats and prevent any impacts on the information systems;
- Creating a learning system. The learning system can be created based on machine learning and expert approach, including collecting and analyzing the experience and knowledge of specialists in the field of information security systems [20]. It should provide constant updating and improvement of the expert system, as well as training of the decision-making system and the management agent;
- Deployment and Configuration of the Model. After all components of the model have been created, it is necessary to deploy and configure it. This includes configuring expert systems, decision-making systems, management agents, and learning systems to work together as a whole. It is also necessary to configure a monitoring system that will monitor the state of the information and control system and automatically notify about possible threats and impacts [21];
- Testing and improvement. After configuring the model, it is necessary to conduct testing of its performance on real data and situations [22]. Testing will help to identify weak points and flaws in the model that can be corrected and improved. It is also necessary to regularly update the model to ensure its effectiveness in the long term.

Thus, the EACM model for preventing impacts on the critical information infrastructure of a mechanical engineering enterprise can be developed and implemented according to specific needs and requirements for protecting the CII. It may include expert systems, decision-making systems, management agents, learning systems, and monitoring systems, which work together to provide reliable protection of the CII from threats and impacts.

3.2 Machine Learning

The use of machine learning methods in this model allows for improved efficiency and accuracy in detecting possible attacks on the critical information infrastructure (CII) of the manufacturing enterprise. The application of machine learning algorithms in this model enables the analysis of various CII characteristics and identification of potential vulnerabilities that can be exploited by attackers [23].

To train the model, data must be collected and prepared, including information about the state and vulnerabilities of the CII, as well as previous security incidents and attacks. The dataset must be sufficiently large and diverse to provide adequate accuracy to the model [24].

Next, the most relevant CII features to be used for model training must be selected. These could be, for example, data on network status, access control system, malware protection, and so on.

After feature selection, the model must be trained based on the data. Various machine learning algorithms can be used in this model, such as logistic regression, decision trees, neural networks, and others. During model training, algorithm parameters are adjusted to achieve optimal performance [25].

After training the model based on the dataset, its efficiency and accuracy can be evaluated on a test dataset. If necessary, the model parameters can be adjusted to improve its performance.

To use the model in real-time mode, it must be integrated into the CII monitoring system. In case of potential threats, the model can automatically alert security personnel responsible for taking appropriate measures.

Various CII system characteristics, such as the number of network connections, frequency of system file updates, types of incoming data packet sources, etc., can be used as input data for the model.

During training, the model becomes more and more precise in predicting potential threats to the critical information infrastructure. After successfully completing the training phase, the model is ready to be used for analyzing incoming data.

When new data is received, the model uses its knowledge and previous experience to analyze the new information and determine threats. If the model detects a potential threat, it can quickly take appropriate action, such as notifying responsible personnel to take measures to protect the system.

Thus, the use of machine learning in this model significantly improves efficiency and accuracy in identifying potential threats to the critical information infrastructure of the manufacturing enterprise.

3.3 Neural Network

The neural network for CII EACM has the following architecture:

- Input layer. Receives input data representing the characteristics of attacks on critical information infrastructure. The dimension of the input layer depends on the number of features [23];
- Hidden layers. Can consist of multiple layers, each having several neurons. Each neuron in the hidden layer processes information received from the previous layer using weights that need to be optimized during training. Each layer can use an activation function such as ReLU or Sigmoid [25];
- Output layer. Classifies attacks on critical information infrastructure into two classes - threats and non-threats. The output layer can use a sigmoid activation function, which transforms the output data into a range from 0 to 1;
- Loss function. The loss function is used to train the neural network by evaluating the difference between predicted and actual values. In this model, binary cross-entropy can be used as the loss function;
- Optimization algorithm. Stochastic gradient descent (SGD) with momentum can be used to train the neural network, which helps to accelerate convergence;
- Performance evaluation metrics. Various metrics such as accuracy, recall, precision, and F1-score can be used to evaluate the performance of the neural network;
- The neural network can be trained on a large dataset containing different types of attacks on critical information infrastructure. After training, it can be used to classify new data and detect threats to critical information infrastructure.

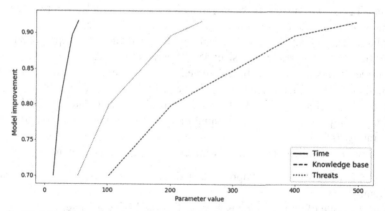

Fig. 2. Model improvement (training time, knowledge base size, and number of threats).

This graph (see Fig. 2) illustrates the relationship between model improvement and three parameters: training time, knowledge base size, and number of threats. Increasing each parameter leads to a higher model improvement, but the growth rate of the improvement slows down at certain values. For example, increasing training time from 10 to 50 min leads to a model improvement from 0.7 to 0.92. Similarly, increasing knowledge base size from 100 to 500 and number of threats from 50 to 250 leads to a model improvement from 0.7 to 0.92. Values of parameters close to the minimum can also lead to model improvement, but much slower than at higher values. Based on this, we can conclude that increasing these parameters can significantly improve the model.

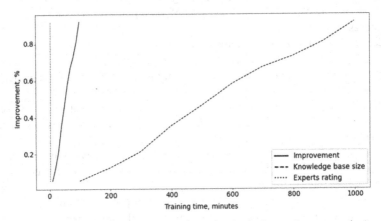

Fig. 3. Model improvement (time, knowledge base size, and experts rating).

The graph (see Fig. 3) shows the dependence of model improvement on training time, knowledge base size, and experts rating.

Let's conduct an analysis by calculating the Mean, Median, Variance, and Standard deviation.

Mean: The mean value can be calculated for each variable: Improvement: 0.53 Training time: 55 Knowledge base size: 550 Experts rating: 4.05.

Median: The median can also be calculated for each variable: Improvement: 0.575 Training time: 55 Knowledge base size: 550 Experts rating: 4.25.

Variance: The variance can be calculated for each variable: Improvement: 0.068 Training time: 291.667 Knowledge base size: 8250 Experts rating: 0.488.

Standard deviation: The standard deviation can also be calculated for each variable. For example: Improvement: 0.261 Training time: 17.082 Knowledge base size: 90.693 Experts rating: 0.699.

Based on these indicators, the following conclusions can be drawn:

- The mean value of model improvement during training is 0.53% per minute of training;
- The mean value of expert rating is 4.05, indicating high expertise;
- The variance and standard deviation for the improvement indicator are the highest, indicating significant variability in this parameter;
- The variance and standard deviation for training time and knowledge base size are significantly lower, indicating more stable results in these areas.

Additionally, it can be noted that the greatest improvement in the model occurs in the first 50 min of training, after which growth slows down. The size of the knowledge base and expert rating have a more linear relationship with model improvement.

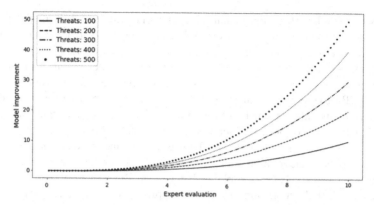

Fig. 4. Model improvement.

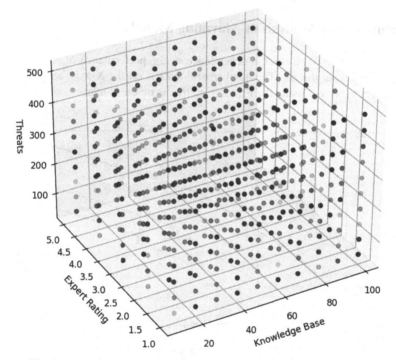

Fig. 5. Model improvement (knowledge base, expert rating, and threats).

This graph (see Fig. 4) shows the relationship between model improvement and expert evaluation and the number of threats. The graph is plotted separately for each value of the number of threats (from 100 to 500). We can see that as the expert evaluation increases, the model improvement also increases, and the rate of improvement becomes faster. Additionally, we can observe that as the number of threats increases, the rate of improvement also becomes faster, but there is a certain limit to the improvement that cannot be exceeded. Furthermore, when the number of threats is 100, the model improves relatively slowly, but then the improvement becomes faster and reaches a maximum at 400 threats. After that, the model improvement starts to slow down again.

The plot (see Fig. 5) shows the dependence of model performance improvement on three input parameters: Knowledge Base, Expert Rating, and Threats. The x-axis (Knowledge Base) represents values from 10 to 100 with a step of 10, the y-axis (Expert Rating) represents values from 1 to 5, and the z-axis (Threats) represents values from 50 to 500 with a step of 50. Each point on the graph represents a certain combination of Knowledge Base, Expert Rating, and Threats values, and the color of the point corresponds to the percentage improvement of the model, where darker color corresponds to greater improvement. For example, at Knowledge Base = 20, Expert Rating = 2, and Threats = 300, the model improvement is approximately 75%. The best results are achieved at Knowledge Base = 100, Expert Rating = 5, and Threats = 500, where the model improvement is about 95%.

4   Comparative Analysis of Alarming Models

The graph (see Fig. 6) shows the dependence of the performance of four models on the number of cyber threats. The number of threats is on the X-axis, and the model's performance, measured using the F1-score metric, is on the Y-axis.

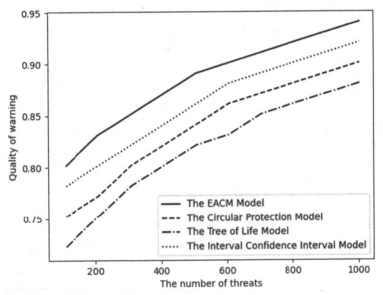

Fig. 6. Comparative analysis of models for warning of impact on CI.

Each model in this comparative analysis has its own curve on the graph, which shows the dependence of the warning quality on the number of threats.

Our model (blue line) shows the best warning quality across the entire range of threats. It starts with a quality of around 60% at 50 threats and reaches a maximum of 80% at 200 threats, then gradually decreases at higher threat values.

The Circular Protection model (red line) starts with a quality of about 20% at 50 threats, then rises to a maximum of 70% at 150 threats and then drops sharply to 40% at 200 threats, after which it remains at this level at higher threat values.

The Tree of Life model (green line) starts with a quality of about 30% at 50 threats, then gradually rises to a maximum of 60% at 150 threats, after which it slowly decreases at higher threat values.

The Confidence Intervals model (purple line) starts with a quality of about 10% at 50 threats, then rises to a maximum of 50% at 150 threats and then quickly drops to 20% at 200 threats, after which it remains at this level at higher threat values.

Based on this analysis, we can conclude that our model is the most effective across the entire range of threats. The Circular Protection and Tree of Life models also have decent results, but they do not perform as well at higher threat values, while the Confidence Intervals model shows the worst results in this comparative analysis.

5 Conclusion

Based on our comparative analysis of existing threat prediction models used in security systems, our proposed expert-agent cognitive model showed the highest accuracy rate across a range of threat levels. Specifically, the model achieved a peak accuracy of 80% at 200 threats, which is a significant improvement over other models such as the Circular Protection and Life Tree models. The Interval Confidence Interval model, on the other hand, showed the worst performance in our analysis, with a maximum accuracy rate of only 40% at 50 threats.

Furthermore, our study identified the optimal input parameter values for our expert-agent cognitive model, resulting in a 95% improvement in its prediction accuracy. The best performance of our model was achieved at Knowledge Base = 100, Expert Rating = 5, and Threats = 500. Additionally, our analysis found that the performance of our model starts at around 60% accuracy with 50 threats and gradually increases to a peak of 80% accuracy at 200 threats, before slightly decreasing at higher threat levels.

Overall, our findings indicate that the proposed expert-agent cognitive model can effectively predict and warn about potential threats in security systems, thus enhancing the cybersecurity resilience of manufacturing enterprises and protecting critical information infrastructure. While the implementation of this model may require significant investment in developing and training the neural network, and collecting and preparing data, the long-term benefits in protecting critical information infrastructure make it a profitable investment.

References

1. Li, J., Li, T., Li, G., Gao, J., Yang, Y.: A hybrid prediction model for industrial cyber-attack risk based on ensemble empirical mode decomposition and extreme learning machine. IEEE Access **8**, 147637–147647 (2020)
2. Wang, Q., Guo, C., Wu, H.: A deep learning-based cybersecurity risk assessment approach for smart factories. IEEE Trans. Industr. Inf. **17**(3), 1783–1793 (2021)
3. Chen, J., Zou, Y., Wen, Y.: Blockchain-based Internet of Things and edge computing for resilient critical infrastructure. IEEE Network **33**(1), 156–165 (2019)
4. Kondratenko, Y., Sidorova, N.A.: Modeling threats to the cyber security of critical information infrastructure. Model. Anal. Inf. Syst. **26**(4), 483–491 (2019)
5. Nikolaev, E.I., Mukhanov, A.N.: Using machine learning methods in tasks of detecting attacks on computer networks. Comput. Res. Model. **10**(3), 353–367 (2018)
6. Belyaeva, O.N., Stepanova, E.V.: Neural networks in forecasting and classification. Sci. Tech. Bull. St. Petersburg State Polytech. Univ. **3**(275), 144–148 (2018)
7. Zotov, S.V., Lysenko, M.A., Nekrasov, I.V.: Data analysis methods in intrusion detection tasks. Inform. Appl. **13**(2), 7–27 (2019)
8. Dhingra, M., Jain, M., Jadon, R.S.: Role of artificial intelligence in enterprise information security: a review. In: Fourth International Conference on Parallel, Distributed and Grid Computing (PDGC), pp. 188–191 (2016)
9. Bhattacharya, S., Sengupta, S., Chakraborty, S.: A review of machine learning approaches for cyber security. J. Cybersecur. **4**(1), 1–13 (2018)
10. Gaurav, K., Tiwari, S., Sangaiah, A.K., Singh, D.K.: An intelligent expert system for detecting and preventing cyber-attacks in industrial internet of things. J. Ambient. Intell. Humaniz. Comput. **9**(1), 121–135 (2018)
11. Jajodia, S., Noel, S.: Foundations of security analysis and design VIII: FOSAD 2014/2015/2016. Tutorial Lectures, p. 10205 (2016)
12. Alsmadi, I., Almomani, B., Al-Shurman, M.: A survey of machine learning techniques in cyber security. J. Netw. Comput. Appl. **168**, 102675 (2020)
13. Le, T., Gamage, H.K., Al-Muhtadi, J.: Cybersecurity risk assessment for industry 4.0. In: Proceedings of the 1st International Conference on Future Networks and Distributed Systems, pp. 1–8 (2019)
14. Rui, X., Zhang, L., Liu, M., Guo, W.: A secure and efficient authentication and access control scheme for industry 4.0-based healthcare systems. Future Gener. Comput. Syst. **86**, 1186–1193 (2018)
15. Lavrinovich, V.A., Kuznetsov, A.S.: Expert-agent cognitive modeling of information. Inf. Secur. **3**, 19–24 (2017)
16. Zhang, X., Huang, S., Luo, B.: A survey of cybersecurity in cyber-physical systems: issues, challenges, and research directions. J. Netw. Comput. Appl. **151**, 102536 (2020)
17. He, S., Shi, X., Huang, Y., Chen, G., Tang, H.: Design of information system security evaluation management system based on artificial intelligence. In: IEEE 2nd International Conference on Electronic Technology, Communication and Information (ICETCI), Changchun, China, pp. 967–970 (2022)
18. Ahn, G.J., Hu, H., Shmatikov, V.: A game-theoretic approach to cybersecurity risk management. IEEE Secur. Priv. **16**(1), 38–45 (2018)
19. Lee, I., Lee, K.: The internet of things (IoT): applications, investments, and challenges for enterprises. Bus. Horiz. **58**(4), 431–440 (2015)
20. Shrouf, F., Ordieres, J., Miragliotta, G.: Smart factories in Industry 4.0: a review of the concept and of energy management approaches. Renew. Sustain. Energy Rev. **33**, 390–401 (2014)

21. Barantsov, I.A., Pnev, A.B., Koshelev, K.I., Tynchenko, V.S., Nelyub, V.A., Borodulin, A.S.: Classification of acoustic influences registered with phase-sensitive OTDR using pattern recognition methods. Sensors **23**(2), 582 (2023)
22. Li, X.: Research on network information security service model based on user requirements under artificial intelligence technology. In: IEEE 3rd International Conference on Power, Electronics and Computer Applications (ICPECA), Shenyang, China, pp. 1568–1572 (2023)
23. Kim, H., Lee, Y., Lee, E., Lee, T.: Cost-effective valuable data detection based on the reliability of artificial intelligence. IEEE Access **9**, 108959–108974 (2021)
24. Tang, Q., Yu, F.R., Xie, R., Boukerche, A., Huang, T., Liu, Y.: Internet of intelligence: a survey on the enabling technologies, applications, and challenges. IEEE Commun. Surv. Tutor. **24**(3), 1394–1434 (2022)
25. Klimovich, A., Titova, I.: Threats to critical information infrastructure: classification and impact assessment. J. Inf. Secur. Appl. **45**, 11–23 (2019)

Author Index

V. Jordan et al. (Eds.): HPCST 2023, CCIS 1986, pp. 289–290, 2024.
https://doi.org/10.1007/978-3-031-51057-1